Watching, from the Edge of Extinction

The last bear in Switzerland: female, 116 kilograms (255 pounds), shot on September 1, 1904, in Val Mingèr by J. Bischoff and P. Fried. Courtesy of the Schweizerischer Bund für Naturschutz.

Watching,

FROM THE
EDGE OF
EXTINCTION

Beverly Peterson Stearns
and Stephen C. Stearns

Yale University Press
New Haven & London

Published with assistance from the
Louis Stern Memorial Fund.

Designed by James J. Johnson and set in Stempel
Garamond type by Running Feet Books, Morrisville,
North Carolina.
Printed in the United States of America by Vail-Ballou
Press, Binghamton, New York.

Library of Congress Cataloging-in-Publication Data

Stearns, Beverly Peterson
   Watching, from the edge of extinction / Beverly
Peterson Stearns and Stephen C. Stearns.
      p. cm.
   Includes bibliographical references and index.
   ISBN 0-300-07606-1 (cloth)
   ISBN 0-300-08469-2 (pbk.)

   1. Extinction (Biology)—Case studies.
   2. Endangered species—Case studies.
   3. Nature conservation—Case studies.
   I. Stearns, S. C. (Stephen C.), 1946–. II. Title.
   QH78.S734   1999
   578.68—dc21   98-34087

A catalogue record for this book is available from the
British Library.

The paper in this book meets the guidelines for
permanence and durability of the Committee on
Production Guidelines for Book Longevity of the
Council on Library Resources.

10 9 8 7 6 5 4 3 2

TO

Justin Kaohuhoomahiluainaokohalaakau

AND

Jason Kauahooululaunaheleonakuahiwi

You who may open this book
years from now,
wondering how they went:
remember too the eyes that watched them go,
and the tongues that strove
to articulate the loss.

Contents

PREFACE

EXTINCTION IS REAL, and it lasts forever.

It is happening on exotic islands and in tropical rain forests, but it is also happening in our backyards. What difference does it make? Is it worth worrying about? Isn't extinction a natural phenomenon? Is it a luxury to be concerned about species' disappearing when war, famine, and disease plague our planet?

These are some of the questions we have asked, first of ourselves and then of others. We share here some of the stories gleaned in our interviews, along with what we learned from them. We have tried to keep these accounts clear and straightforward, but many of the stories are far from simple.

Extinction *is* natural—to a point. We have, however, exceeded any natural background level of extinction. Plants and animals are disappearing from the Earth at an alarming rate, leaving us a planet whose richness and diversity pale in comparison with what once existed.

No one is sure just how many species have existed at any given time or how fast they are disappearing. The current best guess is that about seven million species now inhabit the Earth; it is a fairly rough estimate. The rate of disappearance is thought to be seventy to seven hundred species per year, or up to two species per day; that figure is also pretty rough.

Ever since life originated on the planet, species have been disappearing. More than 99 percent of all the species that have ever lived on Earth are now extinct. But how much of the current wave of extinctions is "natural"? According to the fossil record, at least during the long periods between the few catastrophic mass extinctions, species generally persisted for between one and ten million years. If there were ten million species on the planet, a natural rate of extinction might be from one to ten species per year. This can be considered the natural or "background" rate.

We have several ways to estimate the global extinction rate for all organisms. Today, the most conservative estimates suggest that the global extinction rate is ten to a hundred times the "background rate."

It is increasingly difficult to deny either the extinction crisis itself or the reasons for it: indiscriminate hunting and fishing, poaching, habitat destruction, and the introduction, whether planned or accidental, of exotic species, especially diseases and predators. And just as each species is unique, so is each extinction, for—as we have seen in our interviews—the causes for each are varied—some subtle and complex, others obvious and simple.

In many cases, animals and plants slip quietly into oblivion. The response of the general public ranges from indignant protest to apathy. But each extinction has special meaning for those whose work and lives have been closely linked to the species that disappears. Their assessments deserve special attention, for they are more than judgments on scientific importance: they are comments on what is lost to all of us, on the effect of that loss on certain values, and on the legacy each generation leaves to the next. In this book you will hear from a few people who have watched species slip toward, or over, the edge of extinction and who have contemplated both scientific loss and the broader significance of the disappearance.

These chapters recount only a fraction of the stories that could be told; the number of potential chapters on extinctions in progress is staggering. At one point, trying to find a good example of an endangered species in North America, we put out an inquiry on the Internet. Within hours, suggestions started coming in, and, as we went to bed in Europe, the numbers soared: by morning we were depressed by the success of our query. This was how we found out about the

Barton Springs salamander, whose story we chose to include; but there were many other possibilities.

We had neither time nor resources to travel around the world, collecting examples of extinctions from all the continents, and we believe that it was not necessary to do so. Nor did we make a concerted effort to represent conservationists and scientists from a broad spectrum of races, cultures, and ages or a balance between men and women. From western Europe, we used telephone, fax, post, and e-mail to make inquiries and then personally interviewed the individuals who appeared to have the most interesting, sincere, and insightful stories to tell. No attempt has been made to present these selections in a comprehensive or systematic (or for that matter random) way, but only to be fair, honest, and clear in our telling of the stories. We report on a small but diverse sample of species, situations, and reasons for extinctions, both global and local. We agree with some leading scientists that local extinctions are just as important as global ones. Their concern is mainly that local extinctions result in a loss of evolutionary potential; we feel additionally that any species that disappears from our surroundings can no longer bring its particular beauty or value to our lives. We have tried to describe accurately the motivations of certain people who are fighting to keep certain species from extinction, in most cases people who have worked over a long period of time with one species or one local population and watched it fade away.

We know that we did not always hear all sides of a given story. Some of the people we wanted to reach were unavailable: developers in Austin, Texas, declined to be interviewed; a key figure in the Serengeti was away while we were in the park; conservationists in Hawaii never returned our telephone calls. We wanted to interview a woman who had demonstrated in Greece against development that would destroy the breeding grounds of loggerhead turtles—but she was in the hospital for the latest in a long series of operations to repair the damage from a beating by her opponents. Faced with the reality of that violent situation, we found ourselves thinking, like Roger Burrows, that we were very naïve. We also decided to have a lawyer check the finished manuscript.

We have written this book not only as a tribute to those species which are on the edge of extinction but for us humans who will be left behind, and in hopes we can draw some lesson from what is happening. The fates of some of these species have not been sealed—not quite. We can still make a difference.

We wanted to show how humans who watch and work with endangered species deal with what must appear to many people as the inevitability of extinction, in a process which, in many cases, consumes their lives. We didn't want just to recite cold scientific facts, for there is much more here that touches us all. These are stories of love and passion, dedication and wisdom, life and death. They are also stories of principled people encountering politics, greed, corruption, folly, and hypocrisy.

Many years ago when we moved to our home in Switzerland near the edge of a forest, one spring evening I heard the clear, unmistakable call of the cuckoo. I froze, listened, and waited. It did not come again. How did I know that it was a cuckoo, a bird now rare in our region and one that I had never seen or heard before? I knew because it sounded so much *like a clock*. That realization stunned, then saddened me, and made me wonder what would be left for our children and grandchildren when they reach my age. Will they know animals like the jaguar or the impala only from names of cars? Will they ever see the live animals that have become national emblems, like the American bald eagle or the Guatemalan quetzal? Will they realize that pumas are more than running shoes? Will they ever see a flight of lovebirds turn and wheel against the sun? Will they ever feel the power embodied in a cheetah racing across the golden savanna or the casual strength of the brown bear, as it lumbers over the tundra, or marvel at the grace of a turtle as it glides effortlessly through a blue-green sea? Will they ever stand breathless with astonishment, as I recently did, at the discovery of hundreds of lady's slipper orchids beside a forest path? I hope so, for these are all precious, beautiful, and links to Nature and whatever name one gives to the power that created it, in any case something far beyond ourselves. It is these kinds of reflections on the uniqueness of the other inhabitants of our planet that should give us pause, to contemplate our place and impact on this earth.

—BEVERLY PETERSON STEARNS

ONE MORNING, when I was in first grade at Kohala Elementary School on the north end of the Island of Hawaii, my father came into the school and asked permission of the teacher to take me out for a few hours. It was an unusual request, but she granted it, for he was her husband's boss. Most men on the sugar plantation would never have dared to make the request in the first place. We got into the jeep, drove down to Upolu Point, and walked over to the edge of the sea cliff. Dad pointed out into Alenuihaha Channel and said, "I wanted you to see this. You'll probably never see it again." It was the winter of 1952–53, the end of the Korean War. He had been in a meeting that morning with Harry Taylor, and I could smell cigar smoke on his khaki clothes. I was barefoot, wearing blue jeans and a white t-shirt. He put his arm around me as we looked out into the channel, with the trade wind whipping up whitecaps and blowing strong in our faces.

Before our eyes were hundreds, perhaps thousands, of whales— big ones, little ones, mothers, babies. Spouting and breaching, they were moving slowly from left to right around the point and down the coast toward Honokaa. Dad's arm was around me; the sun was hot but the breeze was cool.

During World War II, when no organized whaling had taken place, stocks had recovered a bit, and the populations from both the Arctic and the Antarctic had made their annual migration to Hawaii to give birth in a region where the water was warmer and where there were no killer whales and fewer sharks than in the feeding grounds nearer the poles.

Dad and I watched in silence for about half an hour, then went home to lunch at the plantation house in Hawi. Many years later, not long before he died, I asked him whether this incident had really happened, because I hardly credited my memory of it. He confirmed that it had, and that I was not exaggerating the numbers.

I shall not be able to show my boys so many whales in my lifetime. Perhaps their grandchildren will see them again.

—STEPHEN C. STEARNS

Acknowledgments

WE OWE our greatest thanks to the people we interviewed. They gave us their time and shared some of their deepest insights. We were deeply impressed by our first interview, with Christophe and Hedwige Boesch, and wondered how many other people had dedicated their lives to what could become a heartbreaking effort. Some interviews dug deep into the personal feelings of people whom, in most cases, we were meeting for the first time. We know it was not easy—for them or for us. But we *all* knew it was important to write these stories.

Others offered us encouragement, inspiration, and very practical help. This book might not have gotten off the ground had Günter Wagner not taken a draft of the first chapters to read on a long airplane flight and shown it to Jean Thomson Black at Yale University Press. Lorene Simms dusted off her pens and sketchbooks and contributed her skills as an artist, sending faxes and queries from her home inside Yosemite National Park. Wallis Menozzi traveled from Parma to share experiences with her own book and encourage our efforts in the midst of a bleak winter. Tijs Goldschmidt visited unexpectedly, lending support to our endeavor and sharing his poignant insights into human emotions, science, and art. We owe special thanks to Cynthia Baer who, on short notice, took time to go over the entire manuscript and give us much needed feedback and criticism, delivered in her inimitable style. Our manuscript editor, Susan Abel, earned our gratitude for her perseverance in correcting details and improving our grammar as well as our message—so that we said what we indeed thought we had.

We came to realize in writing this book that sometimes people who are poorly prepared for the job of protecting endangered species end up casting the final vote on their future. We are human and make "honest mistakes," but when those with authority over an endangered species make mistakes, an entire species can vanish from the Earth. Cynthia Salley saw this happening with the Hawaiian crow and shut the gate to her ranch to keep out scientific researchers. She was sued by the National Audubon Society and criticized in the local press. Roger Burrows saw it happening and voiced an unpopular hypothesis to explain why the wild dogs of the Serengeti had disappeared. He was attacked professionally and personally, and his permission to work in the park was withdrawn. Mike Hadfield saw this as *Euglandina* was introduced to the Pacific Islands, and he "screamed to the skies to stop them" but was ignored. All of them spoke out bravely and should have been commended for their resolute actions. We owe them a vote of thanks for not remaining silent.

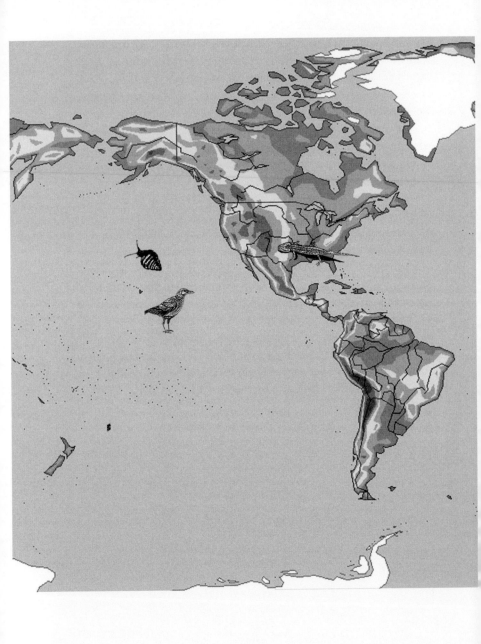

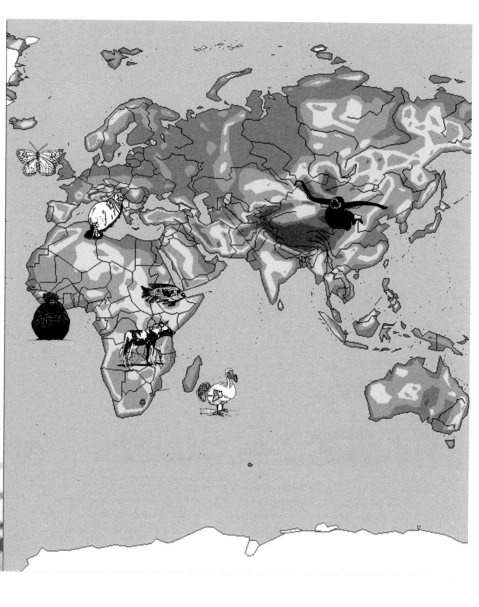

Skull of an 'Alala, *Corvus hawaiiensis,* in the Smithsonian Museum of Natural History in Washington, D.C. The twelve 'Alala in the wild today are all that remain of at least four species of crows that once lived in Hawaii. Photograph taken by Beverly Stearns at the Smithsonian Museum, Washington, D.C.

I
EXTINCTIONS IN PERSPECTIVE

One way of looking at it is that on islands the crisis is over. Most of the diversity in birds that existed on islands only a few hundreds or thousands of years ago is long gone. The total number of species lost from islands would probably be equivalent to about 25 percent of total global avian diversity.—STORRS OLSON

Then the bird moved on, and I rewound the tape just to see if I'd got anything or not. I turned it on again and—that bird came back right away. It was the first thing it had heard other than itself in a long, long time. —JIM JACOBI

 IT IS NOT a coincidence that two of the chapters in this book are set in Hawaii and that this first chapter also refers to those islands. This is not just because Hawaii has an extraordinarily large proportion of endangered species, or because—as one reviewer of our manuscript surmised—we went on vacation in the islands and wanted an excuse to stay longer and write about them.

Our feelings for Hawaii run deep because the islands are entwined with our lives. Steve was born in Kapa'au, on the north tip of the Island of Hawaii, grew up on Kohala Sugar Plantation, and after college on the mainland returned to Honolulu. There he met Bev, who was working as a journalist. We married in Kohala, and our sons were christened by Reverend Abraham Akaka with both American and Hawaiian names, the latter given them by an old Hawaiian friend. Steve did his doctoral research on the rapid evolution of introduced fish on Maui and Hawaii. Although for many years we have not lived in Hawaii, we return often to visit relatives and friends, walk familiar trails, and swim off the Kohala Coast.

Because so much about Hawaii—its history and culture, its animals and plants—is bound up with our lives, we know we may not be objective about the changes that we have seen sweeping over the islands, which have led to many extinctions of both species and ways of life. This "progress" is just as ruthless as the lava that we have watched destroy black sand beaches, *heiaus,* and homes, but the difference is that human beings control this progress—or fail to.

Hawaii has been hard hit by extinctions. It provides a convincing argument that the earth is losing diversity very quickly. But some argue that extinction is just a natural part of evolution. Perhaps, we thought, we see it more vividly in Hawaii, feel it more intensely, because we have watched it happen in a place that is precious to us.

To gain perspective, we traveled to Washington, D.C., to visit Storrs Olson and his wife, Helen James, at the Smithsonian Institution's National Museum of Natural History. We knew that they, as leading avian paleontologists, had a broader view of extinctions than we did, that they could see the current situation against a background of tens of thousands of years. In addition, we knew they had done extensive work on Hawaii's native birds.

In Hawaii, we talked with Jim Jacobi, head of the Hawaii Research Station, one of the field stations of the National Biological Survey. We wanted to know if the situation in the islands was as dire as we thought it was.

Both interviews confirmed our fears: many species have recently been lost, much is disappearing, and the rate of extinction is rapidly accelerating. This is not happening just in Hawaii or only on islands. Take a look around *your* old neighborhood.

In the cramped library of the Smithsonian's Bird Division, far from the museum's public areas, Storrs Olson and Helen James talk quietly about how much diversity was on Earth before humans began to keep written records, how much has been lost, and why.

As a curator in the Smithsonian's Department of Vertebrate Zoology since 1975, Storrs has spent much time in the museum; he and Helen have also spent twenty years literally sifting through the sands of time in the field and reconstructing birds and how they lived—as well as how they disappeared—long before people thought or wrote about natural history. Storrs warns against being naïve about the meaning of extinction.

"I think we have to look at human-caused extinction in terms of some other set of values," he says. "If we are going to take over the earth only for human comfort, then we can manage with fewer species. If we want to learn about what was here, and what evolved, and how it got to be there—and enjoy it for its own sake, then we have to preserve it. I don't think that human development and comfort is necessarily tied in with preserving every species of organism in the world; there's no good evidence for that. And if we maintain that sort of viewpoint or philosophy, it's going to be undermined, because it isn't true. So we should use some other value system, we have to emphasize something other than direct human usefulness and not be disingenuous about what extinction really means."

It is true, he adds in speaking about endangered species, that some plant may offer a cure for cancer, but it is highly improbable that scientists will find such a cure in a snail darter.

Storrs believes if people value the natural world for its own sake and preserve animals and plants because each is unique, then the native character of wild places—forests, fields, shorelines, mountains—can be maintained to some degree. He knows also that "some people would be perfectly happy to have the whole world paved over and be New York City—but there are a lot of other people who don't want to live in that kind of environment."

In Storrs's mind there is no question that people do enjoy nature and knowing about diversity. "Look at dinosaurs," he says, with a nod toward the crowds of tourists milling around one of the most popular exhibits in the museum below. "Look how much enjoyment people get out of contemplating what life must have been like when there were dinosaurs and how diverse they were. But having species only in museums and studied only as extinct things rather than living organisms is not a bright prospect, I think."

He is sympathetic to people who are more worried about needing to eat than about saving species, but he says it is important to balance their position with that of others, who are concerned about extinction because they want the world they live in to be aesthetically pleasing.

"You've got two different forces there that are going to have to be reconciled," he concludes.

How do you think these forces can be reconciled? he is asked.

"I don't know," he says. "I'm not an economist."

He grew up in Florida and, as a child, often accompanied his father, a physical oceanographer who loved to fish. Storrs says his childhood interest in fish, though, was probably due more to his early and fairly constant exposure to marine labs and ichthyologists than to his father. When he was twelve, a local ornithologist took him on a Christmas bird counting expedition, and his loyalties changed forever. He began by studying rails and was led to islands, which are often colonized by rails, and then to the study of fossils, first in the Atlantic and then in the Pacific. With undergraduate and master's degrees in biology from Florida State University in Tallahassee, he went to the Smithsonian to work at the Chesapeake Bay Center. There he met scientists from Johns Hopkins, where he became a graduate student in 1969. He traces his interest in paleontology to curiosity about "systematics and bones and what they tell you about relationships." He studied extinct rails on Ascension Island and then moved further south in the Atlantic to St. Helena, where he found that many birds besides rails had gone extinct. He continued to work on other south Atlantic islands, including Fernando de Noronha and Trindade, then in the West Indies and, finally, in Hawaii.

Storrs's interest in fossils began at a time when people were just becoming aware of extinctions on islands. "We knew about some extinctions on Mauritius, of course, and about the moas in New Zealand, but there really was no idea that extinction had been pervasive throughout the islands of the world," he notes. "Now, of course, that's the viewpoint that's coming across."

In August of 1971 he moved into an office at the Smithsonian's National Museum of Natural History and came into closer contact with Alexander Wetmore, whom he had met in 1967. The relationship proved pivotal. He describes Wetmore as having been one of the world's leading ornithologists from the first part of the twentieth century up until his death in 1978.

Wetmore had led the Tanager Expedition to the mid-Pacific in 1923 and had joined the Smithsonian the next year, becoming first director of the National Zoological Park, later assistant secretary, and then secretary of the Smithsonian. After his retirement in 1952, he continued to work as a research associate. He had published prolifically, particularly on fossil birds. In 1943, just before he launched a study of the birds of Panama, which was to occupy him until his death at the age of ninety-two, he described an extinct goose from the Island of Hawaii. It was the first fossil bird from the Hawaiian Islands to be described and would remain the only one for more than thirty years.

In the early 1970s, Storrs says, the Bishop Museum in Honolulu shipped to Wetmore, in a plaster jacket, a fossil bird still in sand from a dune on the Island of Molokai. Later, bones collected from the dunes of Molokai were sent, as was some material from a cave on Maui. "That sat in Wetmore's office for years, and when he wasn't in I'd go in and look at it and try to identify it," Storrs recalls. At the time, Storrs had never been to Hawaii, didn't know Molokai from Maui, and knew little about the islands' fauna and flora. But drawing on his ornithological experience, he concluded that the bird shipped in the plaster jacket was a *moa-nalo,* a representative of a new species of flightless waterfowl, which they named *Thambetochen chauliodous.* After much consideration, he decided the leg bones from the Molokai dunes were probably from an ibis, a conclusion confirmed when the museum received pieces of bill from the cave specimen on Maui. He wrote up descriptions of both the flightless waterfowl in plaster and the bird from which he believed the bones and bill came and showed them to Wetmore, who at the time was immersed in his magnum opus on the birds of Panama. Together they wrote the manuscript on the Hawaiian discoveries in 1976. The same year, Storrs made his first trip to Hawaii.

That trip marked the beginning of many more fossil bird discoveries in Hawaii. Helen joined Storrs in 1977, and since then they have collected thousands of fossils from Hawaii. Wetmore, Olson, and James have formally described in total thirty-five species of fossil Hawaiian birds, but in Storrs and Helen's 1991 monographs they give informal notice of perhaps twenty-two additional species. The number of known extinct Hawaiian bird species continues to increase as Storrs and Helen write descriptions on a backlog of collected material.

Storrs credits S. Dillon Ripley, secretary of the Smithsonian Institution, for continuous encouragement and support. After Wetmore's death, Ripley asked Storrs for a list of the projects that he had been working on with Wetmore. Storrs listed the projects and what he wanted to do. His memo came back with a simple "OK—SDR" on it, which was enough to ensure funding for him to study the fossil avifauna of Hawaii and to continue other projects. The Hawaiian project has taken much longer than Storrs had anticipated. It's still going strong, and not all the islands have been explored yet.

When Helen looks back at the beginning of their work in Hawaii, she realizes how traditional their view of the fossil record was at first. "Okay, we have different ages of material and different ecological settings—this must explain the differences we see in the faunas," she says, reconstructing the way they thought about it initially. "The passage of time—a great deal of time—allowed these species to become extinct." Then she smiles and adds: "Later, when we were able to get some radiocarbon dates, we had a real epiphany and realized that these birds had become extinct only a few thousand years ago." The time, she calculates, may actually have been only fifteen hundred years, if the usual date quoted for the arrival of the Polynesians, fourteen hundred to sixteen hundred years ago, is used.

"We have no evidence, virtually no evidence, of prehuman extinctions," adds Storrs. He and Helen do, however, have evidence that the destruction of the greater part of the avifauna took place well before Captain Cook arrived in Hawaii in 1778. At least forty-five species of Hawaiian birds, they believe, went extinct before Cook's arrival, an impressive figure considering that there are only about six hundred species of breeding birds in all of North America today.

When Helen was nineteen, she volunteered to work in the Smithsonian paleobiology department. She had grown up in the Ozarks in Arkansas, fascinated with finding arrowheads and other "clues to the past." She describes herself as "a second-generation natural historian"; both her parents are ecologists working on extant avian species. At the University of Arkansas, where she majored in anthropology, she became interested in human osteology and the evolution of hominids. Beginning in 1975, she worked as a volunteer at the Smithsonian and as a research assistant on a project on hummingbird anatomy in the bird division. When the grant for the project ran out, she applied for a job as research assistant working on the Hawaiian fossils that were then coming into the museum. She carefully told Storrs at that time there were two things she would not do: kill birds and work in caves. "And this is very ironic," she says now, "because, in fact, I discovered before long that to do the job of identifying the fossil material, we needed a very good reference collection, and that didn't exist." She helped gather the specimens for the collection that would make the research feasible, and most of her fossil collecting in Hawaii has been done in caves.

Since 1977 the couple have gone back to Hawaii nearly every year, finding fossils in a variety of geological settings—sometimes in dunes, sometimes in lava tubes or pits or fly-in caves in cliffs that only volant birds could have entered. The two have dug into sediment accumulated in karst sinkholes on a weathered limestone reef on Oahu and explored a Pleistocene lake deposit more than 120,000 years old.

In the early days of working in Hawaii, they averaged three new species per trip; today the number has dropped to one. But as those numbers have dropped, the questions have piled up—for instance, just how many island extinctions were there? "You know," Storrs says, "if a species becomes extinct on Maui and on Molokai and on Oahu, there are really three separate extinctions of the same species. There is only one species, but the total number of island populations is another figure that you need to be able to generate."

They point out, for instance, that the 'Alala or Hawaiian crow (*Corvus hawaiiensis*), today found only on the Big Island, was one of at least four species of crows in the islands and the smallest of all. They have found fossils of two other crows (*C. impluviatus* and *C. viriosus*) on Oahu and Molokai. Still other crow bones have been found in lava tubes on Maui and the Big Island, but they need further study before the species can be described and named.

Storrs estimates that two to three times as many species probably formed the modern biota of Hawaii as were reported by Europeans after they arrived. "And this is true for all islands where you have a record in the Pacific," he adds. "There has been 50 to 80 percent extinction, or more."

At the far end of the extinction spectrum is Easter Island, where virtually 100 percent of the native ecosystem disappeared, primarily owing to early man's activities. Lying twenty-three hundred miles west of Chile, Easter Island was discovered about A.D. 400 by Polynesians, who cleared the forest to plant crops and used the timber for canoes and rollers to transport huge statues. They achieved complete deforestation; the soil eroded, and fishing dwindled for lack of canoes. Easter Island could no longer support a population that had risen to seven thousand by A.D. 1500. The civilization that had once carved, transported, and erected the formidable stone statues collapsed into war and cannibalism, and scattered groups of people were left on a barren grassland littered with fallen idols. An expedition dispatched by the Spanish viceroy of Peru in 1770 reported a population of between nine hundred and three thousand people; when Cook arrived four years later, after what may have been a civil war, he found a poverty-stricken population of fewer than a thousand. The statues were no longer worshipped; they appeared to have been toppled by the inhabitants. By 1860 the population had recovered to nearly three thousand, but then it was decimated by a slave raid from Peru and by the introduction of smallpox (*Encyclopaedia Britannica*, 15th ed.).

Storrs acknowledges that while many native species disappeared throughout Polynesia, most of the people did just fine. "They had their taro and sweet potatoes and bananas, and the fact that the birds and snails and so forth went extinct made very little difference," he says. But the Easter Island experience was another matter entirely and should give us something to think about, Storrs notes. "When you get to the point where you've totally eliminated everything in your environment, as on Easter Island, and people are killing each other and don't have enough to eat, you can see that's not conducive to a nice human environment."

He cites fossil evidence of numerous extinctions on several islands in the Pacific. From some of the larger areas, Samoa and Fiji for example, little information is available. "That's because nobody's done the work; the stuff is there to be found, I'm sure," he says. "The literature is growing at such a rate that even the specialists can't keep up with it."

Helen mentions a 1993 review by P. Milberg and T. Tyberg, "Naïve

Birds and Noble Savages," which lists about two hundred species of island birds that became extinct in prehistoric times.

More than half the terrestrial bird species in the Pacific Basin were lost. Even more impressive was the *number* of seabirds that died, far greater than the number of terrestrial birds. Storrs believes few people have realized the extent of human impact on seabird populations—or the birds' leading role in seafarers' discoveries. For him, there is little mystery in the Polynesians' success at navigation: voyagers simply followed the masses of migratory birds, in the knowledge that they would lead the way to both land and dinner.

The commonest fossils in Hawaii are bones of seabirds, especially Dark-rumped Petrels, adds Helen. "There are just zillions of these bones, and now the Dark-rumped Petrel is barely hanging on." She speculates that the elimination of huge numbers of the birds may actually have impeded human navigation and communication in the Pacific, as well as removed an important source of protein for the Polynesians. Very few species of seabirds were lost entirely, she says; rather, "their populations were diminished by several orders of magnitude."

Storrs looks beyond the vanished seabirds to the impact that their loss has had on ecosystems. "This has to be affecting marine ecosystems," he states. When surface predators that feed on millions, perhaps billions, of tons of squid and fish are removed from the marine ecosystem, what happens to the nitrogen cycle on the islands? "We're talking about major alterations of the whole ecosystem just by eliminating seabirds alone, even if you never cut a tree." When the birds disappeared, the squid and fish on which they had fed became available to other predators living in the ocean.

Helen tells of attending a workshop with scientists who were studying how the Hawaiian ecosystem functions. They had noticed that the native flora—especially the nitrogen-fixing plants, of which there are very few in Hawaii—has a hard time colonizing new lava flows. Once the nitrogen-fixers get established on the flows, however, they thrive and pave the way for other plants to colonize. Her colleagues asked her whether the extinct birds could have been bringing the seeds of nitrogen-fixers out onto the new flows.

"The first thing I told them was that there were probably a lot of Dark-rumped Petrels breeding on those new flows," she recalls. "Petrels bring a tremendous quantity of marine nutrients onto land during their long breeding cycle," she explained. Their droppings, and the droppings of their chicks, supplied a steady flow of fertilizer from sea to land. "If you had the petrels bringing nitrogen, you might not need the nitrogen-fixing plants to get out there before other types of plants."

The earlier abundance of Dark-rumped Petrels on the Island of Hawaii was confirmed by the discovery of lava bubbles and caves full of their bones far inland near the Saddle Road, between Mauna Kea and Mauna Loa. Here, Helen believes, were real "petrel-processing factories." She suggests that the petrels were a major food source for the Polynesians, who found the huge colonies. The petrels, she explains, cared for a single chick in a crevice in a rock. The chicks, which remained in the nest for months and grew to twice adult weight in fat, were delicacies there for the picking. The Polynesians eventually reserved the young birds for royalty, but Helen cites fossil evidence that even before Europeans arrived with the rat and the mongoose, colonies of the Dark-rumped Petrel had disappeared.

Early Hawaiians also collected birds whose feathers they used to make beautiful capes for royalty as well as helmets, *kahili* (standards), and leis. It has been estimated that feathers from about eighty thousand *Mamo* (*Drepanis pacifica*), one of Hawaii's unique honeycreepers, were used in the yellow feather cloak of Kamehameha I. It is not surprising that the *Mamo* was nearly extinct by the late 1890s, about the time the monarchy came to an end. The *O'o*, a predominantly dark Hawaiian honeyeater with two small wing-tufts of yellow feathers, was luckier—for a while. Although its yellow feathers were also greatly valued for capes, it was said that they could be extracted without much harm to the bird. Still, it is painful to calculate how many *O'o* may have contributed to one yellow Hawaiian feather cape. Even though Kamehameha is reputed to have said that the birds' feathers belonged to him but the birds themselves belonged to his heirs and should not be harmed, many birds may never in fact have been released after feathers were removed. Stories, and recipes, describe the birds as delicious.

In addition to hunting, habitat destruction and disease have taken a great toll on Hawaiian animals. "Various old collectors reported finding birds dead and dying, which is very unusual," Storrs says. "You have to have a lot of birds dying to find any at all."

Helen has seen evidence of pox in specimens of some of the birds from the late nineteenth century, after the disease had arrived with introduced birds. Avian malaria spread because mosquitoes were introduced in 1827 at Lahaina when American whalers dumped stale drinking water out of barrels loaded in Mexico. Until the mosquito arrived, avian malaria was not a problem, because it had no vector. In addition, birds arrived in Hawaii then, as they do now, as accidentals or on regular migration, and occasionally they carry pathogen hitchhikers.

With a story about Laysan Island, Storrs illustrates how hard it is to sort out causes for extinctions. A man named Max Schlemmer brought rabbits to Laysan, and they wiped out the vegetation. When the Tanager Expedition arrived in 1923, its members found three Laysan honeycreepers that had survived despite the lack of vegetation. But without vegetation they could not hide their nests from egg predators such as Bristle-thighed Curlews and the turnstones, with which they had always shared their island. Then a sandstorm came through, blowing away the last three honeycreepers. "So, what's the cause of extinction?" Storrs asks. "The sandstorm? The Bristle-thighed Curlew? The rabbits? Or Max Schlemmer, who put the rabbits on the island? There's always a combination of factors."

Helen adds: "To me the big question is not what caused the extinction but what caused the rarity that led to it. What was the first thing that made the species vulnerable to all the random things that can come along and wipe out a rare species but can't wipe out a common species?"

Both Helen and Storrs have tried to imagine what Hawaii would have looked like before the Polynesians arrived and the extinctions started. In fact, certain isolated oceanic islands had never been inhabited by humans until Europeans discovered them, and the records of the early explorers, although fragmentary, give us a picture of what such extraordinary ecosystems looked like when they were intact. One such island was Mauritius. The Portuguese discovered Mauritius in 1511 and named the uninhabited and densely forested island Island of the Swans. The Portuguese did not settle there, however, and it was not until the Dutch arrived in 1598 that first reports of flora and fauna were made.

Cécile Mourer-Chauviré, a French colleague who was visiting Storrs and Helen when we spoke with them, has a very good idea of what Mauritius looked like in 1511, for she has studied fossil birds from all three Mascarene Islands—Mauritius, Réunion, and Rodrigues—east of Madagascar in the Indian Ocean. Cécile, the only scientist in France employed to study fossil birds, is overwhelmed with specimens sent to her by French scientists doing excavations all over the world. She says she chose the field after doing a thesis on fossil mammals; she was told that if she switched to studying birds, she could get a job, because it was then a totally neglected field in France. She made the transition, got a job in 1961 with the Centre National de la Recherche Scientifique (CNRS), later spent six years with her husband in Cambodia, and returned in 1971 to the CNRS in Lyon, where she has become director of research.

Cécile says that the Dutch who landed on Mauritius reported "hundreds and hundreds and hundreds of large land tortoises coming down the slopes, and they killed these land tortoises down to the last one." Seeking to stock their ships for the voyages ahead, they patently ignored suggestions that some tortoises should be spared. "And afterward," Cécile says, "there was not one tortoise which remained on these three islands."

She continues: "It was the same for the birds. They have been exterminated by man. I am absolutely sure there was no other cause of extinction." Sketches made at the time depict sailors gathering the tame birds in quantity. A Dutch captain reported: "We found a huge number of turtledoves and other birds. Because the island was not inhabited by people, the birds did not fear us and sat still, so that we could strike them dead without any trouble. In brief, it is a land rich in fish and birds, which are more abundant than we saw anywhere else on our voyage."

Cécile thinks humans also caused extinctions on the Balearic Islands in the Mediterranean, where at least three mammals disappeared about the time that humans arrived. The dwarf hippo of Cyprus was also probably exterminated by human hunters near the end of the Pleistocene.

The discussion extended to animals throughout the world—many of them strange and beautiful—that disappeared before their existence could be recorded, but after the arrival of human predators. One of these was the Réunion solitaire, which was thought to be a pigeonlike bird related to the dodo. Cécile reports that the Réunion solitaire was, in fact, an ibis.

Helen relates that Madagascar was colonized about the same time as Hawaii and New Zealand, and extinctions occurred soon afterward. She reels off a list: "Elephant birds, many large lemurs—lemurs the size of female gorillas—big tortoises, pygmy hippos, all these were lost." As scientists uncover these fossils, they expand our knowledge of the diversity that once existed. At the same time, current extinction rates are accelerating, pushing more species into the past.

Storrs notes that although at least half the species of birds on islands may have gone extinct before their presence was recorded, it was a selective extinction of almost all the flightless birds and many raptors. "So what you're looking at in terms of island faunas is very skewed in the proportions of representation of different groups."

Helen describes how the extinctions, by removing particular parts from the ecosystem, have produced a very different ecological structure than was there earlier. "The native herbivores are gone, the native predators are gone, in Hawaii almost all the granivores are gone, and you may have lost important pollinators." Dispersal of seeds and fruit has been altered, and as a consequence the forest has been affected. Each species that disappears is linked with others, which will either adapt or die out; the chain of reactions continues to alter the structure of the natural world.

The curious thing about the Hawaiian fossil record, Helen points out, is that anyone who is familiar with the birds there today immediately thinks of the colorful nectar-sipping birds that are increasingly endangered. "These birds are all but absent in the Holocene fossil record," she states. "In the Holocene sites, it's all topsy-turvy. We find mostly granivores, and nectarivores are very rare. And it suggests that there's been not just an extinction event but that there's been a shift in ecological structure, that the nectarivores have somehow benefited—they suffered less extinction and they became superabundant."

Added to this puzzle is that some birds, rare today in one habitat in Hawaii, may no longer live at all in other habitats, even though they commonly occur in those habitats in the fossil record. "You can learn a lot about endangered species from their fossils," says Helen. "People study rare birds and try to conserve them in their current habitats where they're obviously not doing well. Maybe before something happened to make them rare, they were more abundant in other types of habitat." By looking at the fossils, it may be possible to determine under which conditions a species could increase in numbers. She cites one bird that was only recently discovered and is very rare on Maui, the *Po'o Uli* (*Melamprosops phaeosoma*). Today it is found only in deep, wet forest, but its fossil bones are fairly common in fossil sites on the drier side of the island. "It is probably just miserable and wet where it is now," puts in Storrs. "It loves to be on the nice dry side of the island where there is no longer any habitat."

So, they are asked, what perspective do they have on people's increasing concern about biological diversity and the extinction crisis?

"One way of looking at it is that on islands the crisis is over," Storrs calmly replies. "Most of the diversity in birds that existed on islands only a few hundreds or thousands of years ago is long gone. The total number of species lost from islands would probably be equivalent to about 25 percent of total global avian diversity."

"I think we haven't learned nearly as much as we could from these extinctions," Helen adds. "The fact that they occurred mainly under the regime of relatively primitive cultures and relatively low human populations, shows that the dangers should be taken really seriously." She believes that only with better management can future extinctions be avoided, but she adds that the global picture is complicated: it is still not clear exactly what went wrong in some places.

Much new information from this fossil work bears on central evolutionary concepts. Storrs says the evidence generally supports the theory of punctuated equilibrium. "We're not seeing graduated changes, certainly," he says. "We are seeing species evolving on the Big Island, which is a very young island, and they are already very distinctive from whatever the mainland ancestor was."

Another insight is that there have been long periods of stasis, during which species have not changed. "We can go back, over a hundred thousand years on Oahu at a Pleistocene site, and these flightless ducks, the *moa-nalos*, are not changing," Helen says. "They had established their morphology and they stayed the same, even through glacial cycles when there were big climatic changes. They were morphologically exactly the same up until this extinction in the late Holocene."

Work with ancient DNA has shown them, for instance, that duck bones found in dry areas of the Big Island are, in fact, not bones from the endemic Hawaiian *Koloa*, a wetland duck, but are from Laysan Ducks. "This tells us that the Laysan Duck is not a Laysan atoll endemic," says Helen. "Laysan is a little atoll that until the 1880s had never been really touched by humans; it's the only place the duck survived. Fossils show that the Laysan Duck was once a woodland, upland species throughout the main archipelago that had undergone a niche shift, into an unusual habitat for ducks." It would be possible, using this information, to re-establish the species somewhere besides Laysan, where it is extremely vulnerable, she says.

Studying the Hawaiian fossils has sparked the couple's admiration for what Helen terms "the evolutionary engine" that provided the diversity. "You realize that Hawaii, with the world's most isolated, complex terrestrial ecosystems, possessed just as many species of birds as islands near continents like the West Indies, under natural conditions," she remarks.

Helen points out that many of the discoveries "play havoc with some of the biogeographic theories, at least for birds," and Storrs adds that "all the numbers that are being crunched are meaningless." They are referring to the theory of island biogeography, which is based on two observations. One is that the number of species on an island increases with the size of the island; the other is that the number of species on an island decreases with the distance of the island from a continent, which is presumed to be the source of colonizers. From these observations a fairly elaborate theory has been constructed and has been widely applied in conservation biology, especially in the design of national parks and game reserves. The new data on fossil birds suggest that much of the original pattern that gave rise to the theory was based on incomplete evidence. They also point to the missing factor: evolution on islands can be quite rapid; hence, isolated areas, far from being depauperate, can harbor a wealth of species.

Helen doesn't look on the birds they dig out of lava tubes and sand dunes in Hawaii simply as fossils. She envisions them in full feather, part of a bigger picture—the ecosystem that has disappeared in Hawaii. "It's important for people to understand that this is not just a tourist mecca, this is a unique, special place," she says. "There is so much here that's yet to be found—that's what makes it fun. When the whole story comes together, it becomes important not just to the paleontologists but to the schoolchildren."

Their discoveries have stirred broad interest, and, Storrs says, their descriptions of the birds frequently elicit disbelief. When he contemplates what Hawaii must have been like before the arrival of humans, he imagines big, tame, flightless ducks and ibises.

Helen comments that they have enjoyed working with an artist who draws extinct Hawaiian birds as they may have appeared in life. "We can use anatomy to come up with the body form and paleoecology to reconstruct the ecosystem, but the plumage, of course, has to be a guess." They have based the colors for the plumage on those of related birds and have taken into consideration variations that are often seen in present-day island endemics, especially in waterfowl. "Then we also sometimes remind ourselves that birds are beautiful, they're spectacular," she adds enthusiastically. "You shouldn't be too conservative in reconstructing the plumages!"

Storrs nods in agreement. "If someone had drawn a bird," he says, "that looked like the Crested Honeycreeper before that species had been discovered, people would have said that no such bird could possibly exist."

Cécile adds that it is a shame that the dodo, another spectacular bird, went extinct so quickly after its discovery. "It's too bad for humanity to have lost the dodo," she says. "It was so harmless."

First described in 1598 by Dutch seafarers, the dodo was big, flightless, tame, and easily captured by hand—although it was repeatedly reported that its meat was tough even when cooked for a long time. In addition to being eaten by humans, the dodo and its eggs fell prey to the rats and pigs introduced by Europeans. It disappeared within eighty years of being discovered. The ungainly bird so captured the interest and fantasy of Dutch and other artists, however, that it continued to be reproduced in paintings long after it had disappeared from Mauritius. It was given its Latin name, *Raphus cucullatus,* only in 1758, long after it no longer existed. The dodo occupies a sad and important place in history, for it is the first species whose extinction was conceded—in writing—to have been caused by humans.

For Cécile, interest in the dodo and fossils stems from a fascination with creatures that will never be seen again. "It is just intellectual interest," she says modestly, but she is pensive about current trends. Avian paleontology, the focus of her life's work, has fallen from interest in her homeland, where a reorganization of biology is taking place. "Actually I am an endangered species in France," she says; "I have so many students who cannot find a job in research. And I know that when I will retire, they probably will not hire anybody to replace me."

While Storrs and Helen excavate the fossils and attempt to reconstruct the bones of birds that have disappeared in Hawaii, Jim Jacobi keeps track of the rarest survivors and does what he can to protect them. Some of the birds that he has watched have passed into Storrs and Helen's domain.

In 1986 Jim accompanied John Sincock, then a U.S. Fish and Wildlife Service biologist, into the rugged Alakai area on Kauai. John was retiring and wanted to show a few people exactly where to find some very special birds. One of those birds, the O'o a'a (*Moho braccatus*), was extremely rare, the last survivor of four species of the genus *Moho.* All of the species had had patches of bright yellow feathers, greatly prized by Hawaiians for use in feather capes and helmets. The Kauai O'o may have survived longer than the others because it had fewer yellow feathers.

The O'o, in fact, had been thought to be extinct. It had been John who had "rediscovered" a few of the birds in this area. Working with limited funds, Sincock and other biologists had tried various management techniques on site, including nest boxes and predator controls—whatever they could, short of capturing the bird and bringing it into captivity. Both the capture and the removal of the bird from its habitat could kill the bird and are controversial actions to take with species like the O'o. The biologists were not very successful, however, with any technique they used.

Jim recalls that biologists had watched the O'o population dwindle to only one bird that they could predictably find, in one remote area. He and John helicoptered in and hiked for about an hour and a half until they came to a ridge. Here, John told him, if the O'o were still alive, was where they would see it. "And sure enough, we heard the bird in the distance," Jim remembers. They carefully climbed down into the bird's territory and suddenly caught a glimpse of it. Then it flew over, and they got a very good look: one bird, immediately identifiable as a male by its coloration. Jim was excited. Remembering a small tape recorder in his pocket, he took it out, and as the bird came by the next time, turned it on and held it out to tape the call.

"Then the bird moved on and I rewound the tape just to see if I'd got anything or not," he says. "I turned it on again and"—he whistles the call—"that bird came back right away. It was the first thing it had heard other than itself in a long, long time." Jim pauses to let the pathetic reality of the deception sink in: the bird had heard only its own call, the call of perhaps the only O'o left. "My first reaction," Jim remembers, "was, 'This is great! Here's the bird back again.' My second reaction was to break out in a cold sweat: 'Why, it's reacting because there's nothing else around!'"

"It was one of the most sobering experiences I've ever had personally, working in the field, realizing, here is something that doesn't look like it's going to be here tomorrow."

And, indeed, he says, the O'o has not been sighted since. He adds, without much conviction, that there is always a chance someone will "rediscover" it, just as John Sincock once did.

The experience prompted the development of a rare-bird search team, the nucleus of which works out of Jim's office in the Hawaii Volcanoes National Park on the Big Island. "It's an expensive operation to get into areas like the Alakai and some other places," he admits, but he thinks that the expense is justified. If any more individuals are found from rare species thought to be extinct, he says he will "run it up the flagpole"—give it as much publicity as possible in order to highlight the plight of the extremely rare species in the islands.

Conservationists have long recognized the importance of "flagship species," which, because of their broad public appeal, are easier to protect than other species. And by protecting the flagship species, conservationists and scientists hope to extend protection to other species in the same habitat. Probably the premier example of a flagship species is the giant panda.

Growing up in Honolulu, Jim had felt that he was in tune with his surroundings, especially the ocean where he surfed as a teenager. In retrospect he realizes how little he knew about the Hawaiian ecosystem and how little the Honolulu schools did to introduce him to the native plants and animals. At the University of California at Riverside he explored an interest in natural history and community ecology. He was especially eager to do fieldwork and returned to Honolulu to earn a doctorate in plant ecology at the University of Hawaii.

Today he works as a botanist with the National Biological Survey and heads the Hawaii Research Station, one of its field stations. Jim's office focuses on Hawaii but also covers the Pacific area generally.

"The Pacific Island ecosystems are in ecological chaos at this point," Jim tells us in his office in the modest, low wooden buildings which are the park's administrative headquarters. "Hawaii has the greatest proportion of endangered species anywhere in the country—in many cases, in the world." But endangered species, he adds, are symptomatic of a bigger problem, the collapse of entire ecosystems.

"As the ecosystems fall apart, as the matrix gets disturbed, more and more species are going to become endangered," he continues. "In many ways, the very endangered species are no longer real functional components of the system; their appeal to people, myself included, is basically emotional." He knows, too, that if scientists concentrate their efforts on dealing with only the most endangered species, the rest of the ecosystem will continue to unravel, "and the problem just gets bigger and bigger."

The objective of the research he directs is to gain understanding of ecosystem function and "what is causing the pieces to fall apart." Ultimately, the goal is to manage and recover entire ecosystems, along with their component species, many of which are birds.

He explains that before the O'o population vanished, for example, it had been very important, not because the individual species was so vital, but because it was part of a whole avifauna in the ecosystem. Look at the difference, he says, between forest at high elevations, where the bird population is still reasonably intact, and the forest at two thousand feet (six hundred meters), where the habitat looks promising but is, in fact, a desert for native birds. "The real concern is that you've got ecosystem processes that are actually being lost." He adds that without native birds that are specialized to cross-pollinate and disperse seeds, the native habitat matrix will not be able to survive. The natural balance will be thrown off, even though some of the surviving insects and the birds are native.

To illustrate the importance of pollinators, Jim tells a story about the native Hawaiian silversword (*Argyroxiphium sandwicense*), which is found only above six thousand feet on Maui and the Big Island. This silvery plant with its daggerlike leaves has gained a high profile and become symbolic of the Haleakala crater area, where rangers have successfully waged a war against the feral goats that once ate it. The silversword grows for up to twenty years, then develops a stalk three to eight feet high with hundreds of yellow and reddish purple flower heads, which produce thousands of seeds. After blooming once, the plant dies. Jim says that because the plant must get pollen from other silverswords to set viable seed, pollinators are crucial. There are no native Hawaiian ants, and the native moths and bees in Hawaii have no defenses against the introduced Argentine ants that have invaded and devastated parts of Haleakala. "There's a very good chance that despite the management that's led to a major recovery of silverswords in Haleakala—and they're up to many hundreds of thousands of plants now because the browsers, the ungulates, have been eliminated—that may be all for naught if they're not able to outcross." The only solution, he says, may be for people to go around with paintbrushes and take over the role of pollinators.

According to Jim, twenty-nine species of birds in Hawaii are listed as endangered, one is listed as threatened, and three more have been recommended for addition to the Endangered Species List, including the 'Elepaio on Oahu, which until quite recently was considered common.

Even the state bird of Hawaii, the *Nene* (*Branta sandvicensis*), continues to be listed as endangered, and Jim points out that no move has been made to downlist it to threatened. "In terms of captive propagation, some people think the Nene is a success story, but though the Nene has been kept from extinction, it certainly has not recovered, by any means." The Nene, "marginally hanging on" in Haleakala on Maui, is not doing very well on the Big Island, either. "They'll nest, but not enough to replace their population as a whole, so supplementation [from the captive breeding program] is very critical at this stage."

The wild population of Nene was down to about twenty-five in the early 1950s, but captive-breeding projects under the auspices of the U.S. Fish and Wildlife Service in Hawaii and the Wildfowl Trust in Slimbridge, England, proved highly successful; the young were reintroduced into the wild on Maui and Hawaii. Nene have also appeared on Kauai in recent years, where they have become a nuisance on golf courses and near airports. The fossil record indicates that the *Nene* were present on all the islands.

Kauai is the island in Hawaii where the bird population has until recently remained most intact, but now two bird species are being proposed as threatened, the Kauai *'Akepa* and the Kauai Creeper, and six are listed as endangered. Jim remarks that until recently there were no known extinctions of birds on Kauai. A reason often given is the absence of the mongoose, which is present on the other islands. Of the six species of birds considered endangered on Kauai, Jim explains, only one, the Small Kauai Thrush, can be accounted for. The others might be extinct: the *'Akialoa* (last seen in 1965), the *Nukupu'u*, the *O'o*, the *O'u*, and the Large Kauai Thrush.

Different organizations use a variety of rules to judge whether an animal or plant is extinct. The Convention on International Trade in Endangered Species of Wild Fauna and Flora defines an extinct species as one that has not been definitively located in the wild at any time during the past fifty years. Jim waves aside the figure, believing that what is important is whether a species is there or not.

He acknowledges that the terrain in many parts of Hawaii is so rugged that it is difficult to determine whether certain birds still survive. Even at his most optimistic, he cannot ignore that Hawaiian birds "are really in dire straits." He has been watching the *I'iwi* pop-

EXTINCTIONS IN PERSPECTIVE

ulation in the islands collapse; it is virtually gone from Molokai and Oahu and is disappearing from areas on the Big Island where it was once numerous. The I'iwi (*Vestiaria coccinea*) is a bright red honey-creeper with a distinctive downward-curved, salmon-colored bill that looks like a small sickle. It once inhabited all the main Hawaiian islands.

Jim vents his frustration about the regulatory and funding bu-reaucracies, which move so slowly that species become endangered and disappear before the agencies take action. "It's been a very frus-trating experience to watch a species fall off the face of the earth." Sometimes the delay in action comes because it is hard to determine why the species is endangered, other times because there is no money to do anything about it.

"If we had the same number of O'o in an area—if we had two dozen birds in an area, as we did back in the late '60s or early '70s," he says, "there are things we could do now and couldn't do back then." The top priority now is an intensive predator control program. Hawaii has huge numbers of feral cats that prey on birds and their eggs; on the Big Island, feral cats roam in both wet and dry areas, from sea level to ten thousand feet. The cats, Jacobi believes, are di-rectly responsible for the 'Elepaio's disappearance from Oahu.

Recently an intensive disease-screening network has also been de-veloped. Jim relies on genetic and serological techniques as well as on visual inspection to determine the health of the birds and the diseases they may have. Only when there is little hope of saving them other-wise are birds taken out of the wild to be protected. "With plants, at least if you find something rare, you can take some cuttings or come back and get the plant itself," he says. With birds, however, that is harder, and their fragile bodies and specific needs make survival in captivity a gamble. It may be as dangerous to take them out of the wild as to leave them in it.

Avian malaria and avian pox have certainly taken a toll. But, Jim explains, pox may actually be an indigenous disease. Unlike malaria, it doesn't require an introduced insect vector in order to be trans-ferred from one bird to the next. Infected seabirds, birds blown off course, or migrants such as the Golden Plover may have brought pox to the islands. Since human contact, he says, other hosts have come to the islands and pressures have increased on the habitats.

Jim also points out that during the past fifteen years much has been discovered about typical habitat for the Hawaiian birds. He cites Storrs and Helen's fossil evidence showing that the *Palila* (*Loxoides bailleui*) once lived in the lowland area of Oahu. Today it lives high on Mauna Kea on Hawaii, above the six thousand–foot level. The existing population occupies only about 10 percent of its former range. This fossil information, Jim says, has "really turned our picture of the Hawaiian avifauna totally on its ear, in terms of what was here and what is gone and where things were."

Perhaps one of the biggest barriers to protecting Hawaii's native plants and animals is ignorance about endemic species. "Most people in Hawaii don't have a clue about what is really Hawaiian, in a natural sense," Jim says.

Over the years a great many plants and animals have been introduced, some inadvertently but others deliberately, to make money. "I don't think it was that they didn't care that those introductions were going to cause problems," Jim comments. "They didn't realize it, and many of them had the best intentions, saying this was a way to bring more birds into lowlands of Hawaii or to bring in some beautiful plants." He tells about one of the park superintendents in the 1950s who brought in fuchsias and planted them along park roads and by lava tubes, the most heavily visited parts of the park, and showplaces of endemic plants such as the *Ohia-lehua*, tree ferns, and orchids. It is not surprising that both tourists and residents are confused about what is a native plant and what is imported.

"There is a very, very low awareness of what our native ecosystems and native species really are," he concludes. He is encouraged, though, by the increased emphasis in school on Hawaiian culture, which extends to native plants and animals. He is also heartened to see the heightened interest in indigenous species among schoolchildren. During a talk about native birds and plants that Jacobi gave to a seventh and eighth grade class on the Big Island, he showed slides of rare species, some of which lived almost in the backyards of the school children. When it came time for the discussion period, some students were angry that no one had told them earlier about the rare plants.

As he has come to recognize native species, he has also begun to appreciate the uniqueness of the Hawaiian ecosystems. "That is really the selling point for me," he says. "We try at times to make economic arguments or functional arguments, and to me the only real strong functional argument is the one on watersheds: the native vegetation is probably the best adapted to protecting the watersheds." Because the natural ecosystems comprise enough different species to be able to tolerate changes in environmental conditions and buffer the whole network, he adds, their diversity prevents its complete destruction. An introduced monoculture, by comparison, is very vulnerable. "The resilience in there is good," he concludes, "but that's a tough argument to give to a hunter."

And he does, confronting sportsmen again and again to explain why preserving rare endemic birds is more important than hunting deer, sheep, and goats on the Big Island. Local hunters had lobbied for the introduction onto the island of deer, which have been responsible for decimating the native forest on other islands. The dry *Mamane-naio* forest on Mauna Kea, along with the Palila, which lives on *Mamane* seedpods, has suffered greatly from grazing by sheep. In 1981 the U.S. Court of Appeals, 9th Circuit, upheld a federal judge's 1979 ruling that the feral sheep would have to be eliminated from the mountain because their presence threatened the survival of the forest. Then, in 1985, the Sierra Club Legal Defense Fund brought suit in the name of the Palila for removal of the mouflon sheep that still lived on Mauna Kea. The Palila won, a verdict highly unpopular with local hunters.

Turning to the subject of tourism in the islands, Jim regrets that the industry has missed the boat in not dwelling on the uniqueness of Hawaii, its flora and fauna. "Not only has it got some real economic potential for them, but it has some very important conservation ramifications," he says. He does notice some progress. "It ultimately comes down to an emotional connection between yourself and what's around you—whether that's with eucalyptus or with an endemic species like *koa*. My experience is that as more and more people realize that Hawaiian components are unique, they will identify with those too."

As Jim works on projects to protect and conserve, he also savors memories of some very bright moments of success and discovery. Having worked in Hawaii over the years, he knows that it is easy to find undescribed species of insects because there are thousands of endemic species. New plants are harder to find, but one or two are recorded in remote areas each year. In 1973, however, Jim was part of the group that discovered a new *bird,* the Po'o Uli, on Maui. "It just knocked us off our heels," he says, delighted at the discovery of a bird that also represented a new species and a new genus.

Achatinella. Drawing of a Hawaiian land snail by Lorene Simms.

2

EMPTY SHELLS

We'll never get the attention and the money that the red birds do. I get so fed up with the red birds that I can't stand it. Every time that I think they spent six million dollars for saving six condors, I just sit and shake my head. How much is going into pandas while entire faunas of invertebrates are vanishing? It's pretty amazing.—MICHAEL HADFIELD

 IN 1888, when J. T. Gulick, the son of Hawaiian missionaries and himself a missionary and naturalist, shipped a collection of Hawaiian land snails to England, he wrote this description in the accompanying letter: "The collection was made during the years 1851–53, when I visited all the districts of the Island of Oahu in person and accompanied by troops of native assistants, ransacked each valley." Then, after references to the destruction of forests and snails, he concluded: "The collection is therefore not only unique but will always remain unique" (quoted in Hadfield 1986, p. 74). Gulick realized that, in his passion to collect Hawaii's beautiful endemic land snails, he had collected *all* of several variants. He may not have known that he had annihilated entire species, but he was certainly proud that these collections would "always remain unique."

Gulick would later be recognized as one of the leading biologists of the age, the discoverer of intra-island endemism of land snails in Hawaii, and the developer of a theory of speciation that corrected deficiencies in Darwin's theory. He was considered brilliant by many; he corresponded with Charles Darwin and Alfred Russel Wallace about the achatinellas, and often sent collections of shells to England.

But with time and perspective, Gulick has emerged, in the words of a scientist who currently works on Hawaii's snails, as "a major scourge," perhaps a scholarly hero but undoubtedly an environmental villain, responsible for the extinctions of some of the very animals that brought him scientific fame. Unfortunately, being collected to extinction is only one of the problems that has clobbered the endemic land snails of Hawaii.

Michael Hadfield went to Hawaii as a marine biologist in 1968; he started observing land snails in 1973 as an avocation. Hiking on weekends with his wife and a friend who studied land snails, he began to realize how fast the remaining achatinelline snails were disappearing, for each weekend they had to climb higher and higher into the Waianae Mountains of Oahu to find one dwindling species, *Achatinella mustelina.* He was horrified at what he saw happening and, as he dug into museum archives, at what had happened during the past century. When the friend who had launched the project in the Waianaes dropped out of academia, she brought him the box of data she had collected, along with the responsibility to put them into shape for publication. Urged on by conservation biologists, he assembled a paper, published in 1980, and has followed the snails ever since.

The definition of a few scientific terms—family, subfamily, genus, and species—may help. The snails discussed in this chapter all belong to the family Achatinellidae. Within that family are several subfamilies, one of which, the Achatinellinae, is found only on the southern Hawaiian Islands. Within that subfamily is located the genus *Achatinella,* which is found only on the Island of Oahu. The formal Latin term ending in *-idae* always refers to a family; the term ending in *-inae* to a subfamily. Genus and species names have various endings. When the Latin terms are put into English, they become colloquial: *achatinellas* thus refers to a collection of snails, perhaps belonging to several species, all in the genus *Achatinella.*

Mike, now director of the University of Hawaii's Kewalo Basin Marine Laboratory, continues his research on marine invertebrates. But he is also running experiments on land snails at a laboratory on the University of Hawaii's Manoa campus, where he teaches invertebrate zoology and graduate courses in larval biology, metamorphosis, and chemoreception. He hikes on weekends now to check the progress of his field experiments on the land snails.

Even though the present-day villains of the saga are habitat destruction, rats, and an introduced predatory snail, *Euglandina rosea,* Mike has not exonerated Gulick and the other early naturalists.

"Those fools in the 1880s who went shell collecting sometimes collected thousands per day," Mike says incredulously, adding that one can estimate from their collections and records just how many snails once lived on the islands. "They were there by the hundreds of thousands." Native Hawaiians did occasionally make shell leis or necklaces from the tree snails, but they had little impact compared with the naturalists, who scooped the creatures up by the bucketful. Mike says that the shell collecting fever peaked during the 1850s through the 1870s, and collecting continued until the 1920s, "when it probably just became too hard to climb far enough to find them."

Gulick, Mike believes, collected more than anyone else. He refers to the field notes of Gulick, Spaulding, and Baldwin, now in Honolulu's Bishop Museum. "The descriptions are incredible," Hadfield says, referring to a passage in the notes telling how young shell collectors would go to Reverend Gulick's house, where they found casks and boxes filled with shells. "Gulick would take a handful and trade them," Mike says with a mixture of amazement and horror. "He was not very good about keeping his locality data or anything like that." He sighs, and adds: "But he did at least make them famous."

Gulick recorded 44,500 shells in three years. How many went unrecorded is open to speculation. He rode horseback throughout the islands, encouraging rural residents to collect for him, buying shells by the thousands and filling his saddlebags with the purchases. And he was not alone among Hawaii's *kamaaina* families in his collecting passion. Names such as Spaulding, Alexander, Baldwin, and Cooke appear on reports from collecting expeditions of the time. Some outings were social occasions. A note in a 1853 edition of the *Weekly Star*, a journal of Punahou School in Honolulu, mentions a picnic after which "all dispersed about in the woods for the purpose of procuring shells. . . . The number procured that day was over four thousand." The same journal reported a week later: "Last Saturday was rendered famous by a pretty general expedition to the mountains in search of shells. Over (2,000) two-thousand specimens were brought back alive by the hardy adventurers comprising about fourteen species of the genus *Achatinella*."

Oahu was not the only island plundered for its beautiful shells. Mike cites a report by Baldwin of a trip to Molokai, where he collected five thousand shells in 1887, and to the Island of Hawaii the same year where, "in a few minutes" in Kohala, he "collected several hundred specimens, picking them from trees and low bushes as rapidly as one would gather huckleberries from a prolific field."

Mike notes that from the museum collections one can tell that the early naturalists collected the adults with the most beautiful shells. "Year after year they went and took all the adults of the species out of these trees, as far as I can see." He describes some areas, such as the region above Aina Haina on Oahu, as now simply "a desert" for snails. And although he concedes that the area may have been also attacked by the introduced *Euglandina,* the endemic snail population had undoubtedly been reduced to such a low level that the *Euglandina* was merely the coup de grâce.

Mike and his co-workers have brought very few snails into the laboratory to breed. His only achatinelline shell collection was given to him by a Hawaiian woman, "Auntie Malia," who telephoned him after an article about his work appeared in a Honolulu newspaper. Auntie Malia told him she had worked at a small Hawaiian museum and, when it was closed, she had kept the museum's snail collection to protect it. She thought it had been made by Reverend Gulick. "It's an old collection," Mike says as he carefully takes it out. "It doesn't seem to be made by Gulick, but there are Baldwin tags on it. . . . There're not enough data in it to be really scientifically useful, because I don't know where most of these things came from, but it's very good to show some of the variety." He identifies many of the shells, some from Maui, some from Oahu. And of others from Oahu, genus *Achatinella,* he remarks regretfully, "I've never seen anything alive even remotely like that."

The shells are lovely, small and shiny, most no more than an inch long, in shades of white, ivory, yellow-gold, and deep brown. It is not hard to understand the desire to collect such beautiful objects, but Mike Hadfield believes they belong, alive, in their native habitat.

The subject shifts to present-day problems and research. That the entire genus of *Achatinella* is now on the endangered species list is both disturbing and, Mike feels, a great boost to the efforts to protect them. But interest in the snails is sporadic. His hopes were raised in 1996 when the U.S. Postal Service announced it was putting together a page of fifteen postage stamps "to promote awareness of endangered wildlife" and sent a photographer to Hawaii. The photographer

Part of Mike Hadfield's only shell collection, given to him by a Hawaiian woman and probably assembled by missionary-naturalists. Photograph by Beverly Stearns.

spent four days taking pictures of the *Achatinella,* but none of them were used. The Hawaiian monk seal is Hawaii's sole representative among the fifteen endangered species from the United States featured on the stamps. Mike notes that only one of the stamps features an invertebrate, the Schaus swallowtail butterfly. The California condor is probably the best-known animal in the collection, which does not include any plants.

When asked how many species of *Achatinella* there are or were in Hawaii, Mike hesitates. He estimates that when Captain Cook arrived in Hawaii in 1778, probably between eight hundred and a thousand species of endemic, terrestrial snails from eleven families were living in the trees, "an enormous richness of individuals and characters." The current situation is quite different. "I bet we don't have two hundred now, and some good fraction of those disappearances is due to *Euglandina.*"

The genus *Achatinella,* which is endemic to Oahu, he explains, supposedly has forty-one or forty-two nominal species. He takes this figure from a 1912–14 study by Pilsbry and Cooke published in the *Manual of Conchology.* Although Hadfield estimates that there may really have been only about thirty-three, he is certain that recent extinctions have taken place, of *Achatinella bulimoides,* for example.

As the introduced predatory snail *Euglandina* spread over Oahu, Mike watched extinctions take place. Every year he hiked up the same ridge in Oahu's Waianae Mountains, and every year the *Euglandina* had climbed higher up the mountain, until finally they had reached the top. "When we last went up there, we could find only dead shells of anything, *Euglandina* included," he remembers. "They had eaten themselves out of house and home on that mountainside."

Euglandina rosea, native to the southeastern United States, was introduced into Hawaii by the Hawaii Department of Agriculture in about 1955 and then again and again over the following five or six years in an attempt to control another introduced species, the large African snail, *Achatina folica,* which had become a pest. Mike says although international agreements were in place to test the effectiveness of introducing *Euglandina* onto the Marianas Islands, north of Guam, the Hawaii Department of Agriculture decided not to wait for the results of the test and went ahead with their introduction. When the number of African snails declined and then disappeared over the next few years in some areas, the Department of Agriculture claimed success. Of course, they were left with the introduced *Euglandina,* which had attacked not only the African snails but the endemic, nonpredatory Hawaiian snails as well.

The irony, says Hadfield, is that the African snail population would have peaked and collapsed naturally. He cites a book by Alfred Mead (second edition, 1980), which documents enormous population explosions of *Achatina* and ensuing collapses. Mike himself has seen this happen on a smaller scale. Outside the marine laboratory where he works, the abundance of *Achatina* on the hedge of *naupaka,* an indigenous Hawaiian plant, generally made it easy for him to obtain samples whenever he needed one. One day when he went out to collect specimens for a colleague in San Diego, he was astonished. "There was literally not a live *Achatina* anywhere," he remembers, only dead shells by the hundreds. And he was sure that *Euglandina* had played no role in this demise, for the area is not the kind of place *Euglandina* lives, and no *Euglandina* shells were among the hundreds of empty *Achatina* shells he found.

Much to Mike's dismay, and over his objections, the Hawaii Department of Agriculture, encouraged by its "success" in Hawaii, proceeded to advocate the introduction of *Euglandina* throughout the Pacific as a biological control agent against the giant African snail. It was introduced onto Tahiti, Moorea, and, in 1980, with Mike "screaming to the skies to stop them," American Samoa. He shakes his head: "Now almost all the native snails in American Samoa are gone. It probably took ten years."

He also notes that although the African snail did not go higher than six hundred or eight hundred feet at the most, *Euglandina* appears to be able to live almost anywhere. "I've picked up *Euglandina* on Konahuanui, the highest peak in the Koolaus," he says. "We finally brought the snails out of there, *Achatinella fuscobasis,* because *Euglandina* was right up in the *i'e i'e* (an indigenous Hawaiian plant) with them." Although he hasn't been back to that peak since, he supposes that *Achatinella fuscobasis* may be extinct there, surviving only in his laboratory.

In the wild, *Achatinella* live in Hawaii's lush wet areas, primarily, but not solely, on native bushes and trees. Mike says he has seen the snails on introduced vegetation such as guava and silk oak, but they don't persist there; they prefer the endemic *Ohia-lehua* (*Metrosideros polymorpha*), *Papala-kepau* (*Pisonia umbellifera*), or *Hame* (*Antidesma* spp.). He has found the same snails, year after year, on those host trees. That is one reason the mark-and-recapture method he employs to study them is so successful.

Mike's laboratory on the University of Hawaii Manoa campus houses several refrigerator-sized incubators, in which live five or six species of *Achatinella.* Temperatures in the incubators fluctuate from 60°F (16°C) at night to 70°F (21°C) in the daytime, simulating temperatures at twenty-five hundred to three thousand feet (750 to 900 meters) elevation. Lights go on at 6:00 A.M., off at 6:00 P.M., and it "rains" every eight hours. The snails live in terraria on branches of *Ohia-lehua,* and the leaves are changed about twice a week. The snails feed not on the leaves themselves but on microscopic layers of black molds that grow on the leaves, not damaging the plants but moving over them with their radula,"a sort of tongue with tiny little teeth in rows across it." Mike compares their tiny teeth to the grit of fine sandpaper.

Laboratory experiments, in which the snails were fed molds cultured on agar, have been successful; the snails have adapted quickly, and a second generation has been born in the lab. After the construction of two snail houses at the university's Lyon Arboretum, the program for rearing captive snails will be expanded. Use of the snail houses, scheduled for 1997, was delayed by bureaucracy.

In February 1994, Mike began reintroducing the achatinellids back into the field. Eighteen snails born in the laboratory were carefully released in the Waianae Mountains. Their host *Ohia-lehua* tree was covered with a nylon screen cage so that they would not run away before they adjusted to their new home. Hawaiian snails, Mike says, are loyal to their tree and tend not to leave it. "You find the same snails in the same tree year after year after year, unless some major windstorm blows them out of their trees, then they crawl up the first tree they come to," he explains. The cage was removed in April 1995. A month later, a survey showed that only half of the original eighteen snails remained; six shells were found of snails that had died; the other three appeared to have escaped into surrounding vegetation. Of another species of snail that he released on Molokai, 30 percent have survived to maturity.

Still, Mike is not discouraged. Given the snails' capacity for self-fertilization, even a 30 percent survival rate may lead to a successful reintroduction.

Achatinella don't reproduce until they are between four and six years old, and they have very few offspring, perhaps four to six a year, so the rate of population growth is slow. They are born relatively large, however, about 4.5 millimeters, and thus have what Mike calls "a running edge at birth." Still, the death rate is high during those four years of vulnerability before they add even one replacement. He cites a study showing that among a species on Molokai, *Partulina proxima*, the adults needed to survive about nineteen years just to replace themselves.

The snails are hermaphrodites, and the eggs are encapsulated in the uterus. Mike estimates the egg diameter as about nine hundred microns at most, but because the animals are four millimeters or larger at birth, it seems likely that the snails are truly viviparous and that the parents nourish the young.

In some respects, the snails' life histories resemble those of whales; they have to live a long time to reproduce enough to replace themselves. And like whales, they have been hunted and have become increasingly rare. Unlike whales, they have low visibility and, without drawing much attention, can be trampled underfoot, gathered up for collections, or preyed upon by introduced species.

Mike's publications make reference to habitat destruction, a leading cause of the land snails' decimation. He quotes Baldwin, in 1887, as already noting that *Achatinella* were suffering from "the ravages of cattle through our forests" and that "the agencies now threatening these little gems of the forest are the rats and mice, which have become very abundant in the mountain forests, particularly where there are no cattle."

Anyone who has visited Hawaii recently probably would not be surprised to read that habitat destruction is the single most important cause of population decline for any species there. "If there's no forest, there are no forest dwellers, that's got to be the first thing," Mike says. Early Europeans who farmed a valley near the current site of Kaneohe on Oahu reported turning up *Achatinella* shells in the dirt of land the Hawaiians had cleared by burning the coastal forest. "So there were coastal species that were driven to extinction by the Hawaiians, without a doubt," Mike says.

In addition, during the nineteenth century, European rats were cited as predators of the land snails. Mike has been surprised, however, that the traps he has set in areas where rat predation is high have yielded only Polynesian rats, the species that arrived on canoes with the first Hawaiians. As is true of birds, many snails may have disappeared as a result of the Polynesian impact, before Europeans arrived.

Why is he—balancing two research interests, teaching, and administrative duties as director of a marine lab—struggling to keep the land snails of Hawaii from going extinct? What difference would it make to him?

"Oh, boy," he sighs, in answer to a question that probably seems simplistic to him. He lists a whole set of reasons. "One is that it's a biological richness that's simply incredible," he begins. Unlike other biological and evolutionary stories from Hawaii, he says "the snails are better than the *Drosophila* (a genus of fruit flies represented by hundreds of species in Hawaii) and better than the birds because they're doing it all [speciating profusely] so much faster." While the *Drosophila* and many of the birds were found throughout the Hawai-

EMPTY SHELLS

ian chain, the *Achatinellinae* inhabited only the major southern islands of Hawaii; they did not island-hop down the old volcanic chain. "That means . . . the radiation into two hundred species in three or four subfamilies—this incredible richness—has a time span of about two million years." Evolutionarily speaking, that is a very short time for what has happened. And as a picture of radiation, Mike sees it as a "magnificent" scientific resource for understanding processes and rates of evolution. "How could you estimate the value of something like that?" he asks.

"As far as we know, within the *Achatinellidae*, within the genus *Achatinella*, for instance, let's say we have twenty to thirty species, all on Oahu, all probably evolving in less than one million years, and as far as we can tell, it is not adaptive radiation. Adaptive radiation is how people think evolution typically occurs, and it's like 'damnyankee,' it's one word. As far as I know, there is nothing adaptive about the radiation of the *Achatinella* species."

Mike believes the snails have formed separate noninterbreeding entities without adaptive change. They live on the same trees, they eat the same food, and they have very similar life histories, but life histories that are nonresilient. "The snails grow at a snail's pace, mature at a very late age, and reproduce at an incredibly slow rate." He adds that selection to resist predation clearly did not play a part in the evolution of the land snails' life histories.

As for the cultural aspect of the survival of Hawaii's land snails, Mike, as a scientist, is eager to accept whatever assistance the Hawaiian culture—or any other—can give to extending the future of the species he has studied so long. He believes that the residents of Hawaii, the nontourist population, are genuinely concerned about the snails. "There are songs and stories about *pupukanioe*," he explains using the Hawaiian word for the land snails. "The 'singing snails' go way way back. Every good, solid Hawaiian is absolutely convinced that these snails sit up in the trees and sing. And I don't even argue it anymore. They don't sing, but anyway, I wish they did."

He thinks that undoubtedly the native crickets, which are tiny, green, and more difficult to see, do the singing, but that when people look for the source of the song, they see only the larger, more colorful land snails in the same tree.

"The Hawaiians have always known about their beautiful snails, and they put them into songs and stories. I think there is a real connection there, between the Hawaiian heritage and the really healthy, intact Hawaiian forest."

But if a snail's image is better than a slug's, it doesn't begin to approach that of exotic birds. Mike envies the attention—and even more, the money—given to studies on Hawaii's endangered and often spectacular birds. Not only have the land snails fallen victim to Victorian shell collectors and imported predator snails; they now have to compete on the world environmental stage with much more glamorous species, many of which are also native to Hawaii. "We'll never get the attention and the money that the red birds do," he says, referring to the spectacular honeycreepers endemic to the islands. "I get so fed up with the red birds that I can't stand it. Every time that I think they spent six million dollars for saving six condors, I just sit and shake my head. How much is going into pandas while entire faunas of invertebrates are vanishing? It's pretty amazing."

He doesn't understand much of the public's response to extinction crises, now or in the past. In the United States, ordinary citizens have been conscious of the threat of extinctions at least since the disappearance of the passenger pigeon in 1914, but to many people extinction "just doesn't matter." Visiting his family in north-central Washington State in 1994, he found himself baited with comments that the spotted owl really wasn't endangered after all and, besides, how could extinction of a bird, a snail, or a bug compare in importance to jobs for loggers in the area? "Environmentalism is a dirty word up in my home country," he reflects.

Like most scientists, Mike applies for grants to support his research. Most grants are short-term, and most are for relatively small sums of money to pay the salaries of technicians or fieldworkers, usually graduate students. In addition to his research, writing applications for grants to ensure that the work goes on consumes much of his time.

"I have never had anything except one-year funding for the tree snail effort," he observes. "No three-year grant, no two-year grant, no five-year grant, and for years, of course, we had no money at all. Even after NSF [the National Science Foundation] started having grants in conservation biology, we would write the best proposals we could, and they would get very nice ranks except for the person who would say, yeah, well, but this won't apply to anything except Hawaiian tree snails, and so we would get the killing mark." Several times his project has almost died for lack of funds. With his current short-term funding from the state of Hawaii and U.S. Fish and Wildlife, Mike will soon have to think again about spending a month to raise money to keep the project going for one more year.

He focuses now on reintroducing the snails from his captive-breeding program into areas that appear to be perfect habitats but from which all snails have died out. These are generally remote, hard-to-reach areas; if he is lucky, he can get in by helicopter. This type of reintroduction, however, works only in areas surrounded by enough "snail desert" so that *Euglandina* won't be able to eat its way through to reach it.

"So that's my prospect, if we can get enough financial support to keep this effort going for another ten or fifteen years . . ." He pauses, then adds, realistically, "Oh, it's going to have to go forever, you know."

He cites some discouraging moments in the attempt to save Hawaii's endemic species, snails or others. One was a recent development in which the state granted a permit to divert water from the source of a stream on Kauai, a move that would change an entire habitat and destroy the animals living there. "I see things like that, and it makes me think there's no hope," he says. "On the other hand, if I didn't think there was hope, I guess I would just go back to my marine lab and work on the development of some sea slugs, and I could keep busy until I retired."

Mike relates an incident to illustrate just how complex and crazy the situation may become in the struggle to protect certain animals. Between 1994 and 1996, in the project on Molokai, rats drastically reduced the local *Partulina* populations. In a move to control the rats, the Nature Conservancy, which was monitoring the project, put out rat poison. Unfortunately, a wild pig broke into the area and ate the rat poison. Because there is widespread pig-hunting on Molokai, the Nature Conservancy had to go public, announcing that wild pig shot in the area might be poisonous. A huge uproar resulted, and the Nature Conservancy found itself under fire from animal-protection groups.

He is asked what he believes will happen to the snail project when

Mike Hadfield dies? He smiles and concedes that it is a good question: he is almost sixty and increasingly feeling his mortality. Then he begins listing graduate students and former students who he feels fairly confident will stay in Hawaii and carry on his work. "So it may outlast me," he says. "God knows I'm trying to get a second generation going on this."

He admits he has become so engaged that he simply can't quit. "If I really get down in the dark of the night, I am pretty pessimistic, but I don't operate from that pessimism. I operate from some optimism that is around more often." That optimism, he says, stems from the success he has had raising the snails in the lab and seeing their populations grow, and from the knowledge that he will reintroduce them into their native habitat. He speaks enthusiastically about the work a postdoctoral student is doing on the snails' genetics, to check for inbreeding. Much still remains to be discovered about the snails, and he is anxious for it to be done soon. He knows that the situation is getting worse on Oahu; two species he last saw in 1988 are definitely gone. Others may be also. Time will tell.

He values the trips into the mountains, up into the cloud forests, where "life starts all over again and you don't think about what's going on back on campus, who is on what committee," or the next grant to write. When, in addition to that, he says, he finds snails, the excitement is is hard to describe: "ecstatic" probably comes close.

He recounts seeing the species *Achatinella apexfulva* in 1985 in an area where later they discovered the *Ohia* trees slashed down with machetes. Although he and his team searched and searched, they were never able to find the species again and finally gave it up as extinct. Nearly ten years later, while he was on sabbatical at Friday Harbor in Washington State, the e-mail flew frantically back and forth when his crew working in Hawaii rediscovered *Achatinella apexfulva*. It was, says Mike, what friends of his at Berkeley call an Elvis species. He laughs: "And so those things are, at least on a very personal level, part of what keeps you running back to the mountains and looking, writing one more grant proposal."

Corvus hawaiiensis. Drawing of an 'Alala (Hawaiian crow) by Lorene Simms.

3
AS THE HAWAIIAN
CROW FLIES . . .

The 'Alala were right across Kona, one end to another, very south Kona to very north. So they were throughout that whole territory. Well, ten years later, guess what? We had the only ones. And it was the only place the researchers weren't allowed.—CYNTHIA SALLEY

 THE ENTRANCE TO McCandless Ranch on the Island of Hawaii is about twenty miles (thirty-two kilometers) south of Kailua-Kona, as the crow flies. If wild crows—the endemic *'Alala,* Hawaii's only crow—still fly here at all, it is thanks to the determination of one woman, Cynthia Salley, who lives at McCandless. Standing firm against the National and Hawaii Audubon Societies as well as Hawaii wildlife officials, and uncertain about the U.S. Fish and Wildlife Service, all of whom wanted to "protect" the endangered crow through a captive-breeding program, Cynthia shut the ranch gate in 1978 and told biologists and researchers to stay away from the nine remaining wild 'Alala on her property. After all, she said, they could study the other sixty-five to ninety wild crows in the Kona area if they liked, but she wouldn't tolerate any more researchers on her property; they had broken their promises to her and had put the crows at risk. It was clear to her that the 'Alala, which she had known since childhood, were more endangered than helped by the researchers and the "bureaucratic biologists" who sent them. The move was, to say the least, unpopular, but Salley was vindicated: ten years later, the McCandless crows were the only wild 'Alala left.

The story, which is long and complex, is essentially about rights: the rights of the landowner, the state and federal governments, researchers and environmental groups and, perhaps last of all, the endangered animal itself. It is also about the tenacity of one woman who decided it was time to use common sense to protect an endangered species she valued.

Sitting on the lanai of her McCandless Ranch home high above the Kona Coast, Cynthia carefully recounts the course of events and the more recent conditions under which she once again allows biologists onto the ranch. As a child, she spent time at a cabin—at sixty-two hundred feet, reachable only on horseback—enjoying the beauty of the *koa*-forested slopes and the sound of the Hawaiian birds on land which her grandfather had begun accumulating in 1915. Today, the fifty-thousand-acre ranch remains in family hands; the portion she controls, about fifteen thousand acres, is used primarily for raising cattle and for logging *koa*, a native Hawaiian tree.

She describes the Hawaiian crow, *Corvus hawaiiensis*, as the most social of all Hawaiian birds, and, with a wry smile, she tells how the inquisitive bird would come swooping down to check out the sound of a hunter's shot. Now, as when she was a child, she can call and they will come to the sound of her voice: she and the others on the ranch have never had to seek out the birds, the birds have always come to them.

Although researchers had been studying the 'Alala for years, suddenly in the 1970s everyone seemed to want access to McCandless Ranch to study them: the U.S. Forest Service, the National Parks Service, the U.S. Fish and Wildlife Service and state biologists. This enhanced interest was due to increased awareness that the crow had become an endangered species in Hawaii, and to the local publicity for the captive-propagation program for the 'Alala, established in 1976 and supported by the federal government and the state of Hawaii. By that time, Cynthia was a resident owner of the family property, and she regularly granted requests to allow researchers onto the ranch—requests nearly always made during the crows' breeding season. The biologists eagerly came and undertook their "research," a word Cynthia now uses advisedly.

For a while, at least, it was fine with her for these "bureaucratic biologists" to come or send researchers to study the 'Alala. "They did, and they did, and they did, and they did," she says. She uses the term *bureaucratic biologists*, even though she knows they dislike it: "It's very descriptive."

"Nobody ever bothered to go back and see what anyone else had done, nobody ever changed what they were doing, it was the same thing, year after year: you go in during breeding season, you climb to the nest, you count the eggs, you go down, come back, try to see if they're still in the same nest—and very often they weren't because they'd been disturbed—climb up see what's happening or see if the nest has been deserted, climb down and leave."

No one, she says, ever simply came to observe the birds' behavior, their feeding habits, their relationships, and, especially, no one came unless it was breeding season. She felt that their reports ended up in a pile on a desk somewhere or were filed away; she saw no indication that the research had led to anything.

She knew that the researchers climbed the trees, counted the eggs, and sometimes put pesticides around the nests to control the insects—primarily mosquitoes, because avian malaria is a real threat to many birds in Hawaii. She was also aware that they were taking fledglings for the state-run breeding program set up at Pohakuloa, near a military training area on the Big Island. About this time, however, she also noticed a definite decline in the crow population. Concerned that the birds might be abandoning their nests because of the frequent human interference, she started talking with the researchers to find out more about their methods and goals. In 1977, when the U.S. Forest Service called about sending someone in to do another project on the crows at McCandless, she announced she was laying down some guidelines: no climbing of trees, and researchers must stay behind blinds to make their observations.

Some time later, a student doing research for his master's degree came to Cynthia and proudly showed her the results of his study, including a film taken of the birds at McCandless and their eggs in the nests. "I mean, we were looking right into the nests in these films," Cynthia says incredulously. "I asked him, 'How did you get those?' He said, 'Oh, we have cameras up in the trees.' I said, 'How often were you changing the film?' He said, 'Two or three times a week.' And I said, 'You're climbing up trees to do this?'" She was infuriated to learn that her guidelines had been ignored.

Cynthia called the person who had set up the student's research project. "I said, 'That's it, you can't come up anymore, nobody can.' And I didn't differentiate between the good guys and the bad guys. I just said, that's it, no more research out here; you know, I'm going to do my own research." That was 1978. According to Cynthia, at the time studies showed nine 'Alala on her land at McCandless and a total of perhaps ninety in the Kona area.

She lists the other parts of the Kona area of the Big Island, running from north to south, where researchers continued to study the 'Alala, always with permission from the landowners. "The 'Alala were right across Kona, one end to another, very South Kona to very North. So they were throughout that whole territory. Well, ten years later, guess what?" she asks grimly. "We had the only ones. And it was the only place the researchers weren't allowed."

Despite pressure from all sides, she stuck with her decision to keep everybody out. "Common sense was telling me that although the population was declining for whatever reason, human interference was accelerating the downhill slide," she says, "and if these birds were to have a chance, they were going to have to stop doing this." She believes that avian malaria and pox were the main cause of the decline of the 'Alala and that predation by mongooses, rats, and cats contributed to it, but that the McCandless crows were also threatened with being studied to death.

During the ten-year period, she says, she continued to talk with people from the state and federal agencies who would call to ask if they could come onto the ranch to look at the crows. "I never refused to talk to them: I would listen to whatever it was they had to say, whatever, but they never had a better solution." She remained adamant. The requests were always the same: to observe during breeding season or during the period just before that time, when the crows were vocal, when they were most vulnerable.

After she had shut the gate to researchers, Cynthia took a deeper look at what was happening at Pohakuloa, where the state ran its captive-breeding program. The program to breed Nene geese had been outstandingly successful. It had brought the Nene back from the edge of extinction in the early part of the century, to the great relief of those who had named it the state bird of Hawaii. The Nene program at Pohakuloa was augmented by the program to breed the 'Alala in captivity and raise fledglings that were, in one case, captured at McCandless Ranch.

The person in charge of the 'Alala project at Pohakuloa when Cynthia first looked at it, about 1979, was Barbara Lee, who has been both criticized and praised for her deep commitment to the crows and the work which began when she was a volunteer in the mid-1970s. In

1977 Lee had moved to a seventy-five-dollar-a-month job as a biological consultant at Pohakuloa. But the program was struggling; neither Lee nor anyone else could account for the disappearance of two young chicks hatched in 1977—or for the chicks and eggs that disappeared over the next three years. One theory was that the captive parents destroyed and ate their eggs and young; another held that rats got into the cages and carried off the young. Cynthia recognizes Lee's deep emotional involvement with the crows and praises her vast knowledge about the birds. Without a doubt, she says, Barbara Lee knew more about the 'Alala than any of the researchers or bureaucratic biologists. Lee would take the fledglings, first collected in the wild by the State and the National Park Services, and raise them in her bathroom until she felt they were old enough to go into the aviaries. She observed them closely, studied their behavior—not just during breeding season—and was totally committed to keeping the 'Alala from becoming extinct.

Barbara Lee also, however, had definite opinions about how the breeding program was being handled—and she was very vocal. State officials fired her in 1981 and hired an aviculturist to take over the 'Alala project. He soon left the job, as did a series of successors. None had much luck raising the 'Alala in captivity. During this time, three eggs were taken to the Honolulu Zoo, and once they had hatched out, the chicks were returned to Pohakuloa to be raised.

As time went on, Cynthia says, what was happening at Pohakuloa increasingly affected her own perspective and the position she took.

Cynthia sighs. "I first got to know Barbara at Pohakuloa, but I didn't get to know her well and realize the extent of her knowledge until after she'd been fired from that position," she says. As pressure was put on Cynthia to allow access again to the 'Alala on her property, she sought out Barbara Lee to learn what she could about the birds' behavior. Barbara confirmed and bolstered Cynthia's opinion that the crows should be left alone.

Then, in 1984, with the arrival of a behavioral biologist to manage the 'Alala program at Pohakuloa, expectations rose for the captive-breeding program—and so did Cynthia's concern for the crows at McCandless. Dr. Fern Duvall, who cited Konrad Lorenz, the renowned Austrian ethologist, as his mentor, was brought in largely because of his experience with crows. Duvall was hired by the state of

Hawaii as a consultant to the Board of Land and Natural Resources Division of Forestry and Wildlife. He insisted that it was crucial to get birds from the wild population into the breeding program to add genetic diversity to the captive group. Although Cynthia refused to allow Duvall to study or capture crows on McCandless, she felt obligated to allow him and a state wildlife biologist access across McCandless to an enclave of state property that was encircled by the ranch. Once on that piece of state land, she says they played recordings to attract McCandless crows so that they could capture them.

"It really upset me that they were trying to lure the birds over," Cynthia recalls. She was naturally relieved when the attempt, which took place during the crows' vocal period prior to breeding, did not succeed; however, it did little to impress Cynthia with the new head of the 'Alala program. Comments that Duvall made about Barbara Lee and her nonscientific methods angered Cynthia and aroused her distrust.

Duvall had no success with the captive-breeding program during his first year. He blamed U.S. Army exercises in the Pohakuloa area for disturbing the birds, so, at his suggestion and with the agreement of other biologists—and at a cost of six million dollars to taxpayers—the facility and its captive birds were moved in November 1986, to the Island of Maui, to the site of the state's former Olinda Prison Honor Camp. Still the program met with no success; the incubator temperature was blamed. The following year, lack of genetic diversity was cited as the cause.

Under Duvall's tenure, about seventy eggs were laid at Olinda. Only five hatched, and four birds were reared. The unhatched eggs, laid by birds that had been moved from Pohakuloa, were all deemed infertile. No new birds had been taken from the wild.

But Cynthia's greatest concern was over what would happen if the biologists became convinced that the root cause of the 'Alala's reproductive problem was lack of genetic diversity, a concern that Duvall voiced in interviews with the press. The only solution to that problem rested on her land.

"At that point," she says, referring to the years 1987–88, "I really closed down." Her worst fear was that someone would jeopardize the fragile 'Alala population at McCandless to support a program that wasn't working.

Her continued refusal to grant entrance to any more researchers thrust the plight of the endangered 'Alala into the public spotlight and gave rise to newspaper headlines such as "Hawaii Rancher: Hands Off Crows."

Cynthia dug in, confident that her decision had been right. What perhaps cut the deepest were charges, first made to her and then in island newspapers, that she would be to blame if the 'Alala became extinct. "It was a small cadre of biologists, who said that perhaps I meant well, but what I was doing, if the crows went out of existence, it would be my fault; I would be solely responsible."

About this time, during breeding season in May 1989, Cynthia learned that a team of state biologists led by Duvall planned to use a helicopter to capture crows over forested land on McCandless Ranch. Duvall, now a permanent employee of the state, had become its aviculturist in charge of developing the captive-propagation program for not only the 'Alala but all of Hawaii's endangered bird species. One of the state biologists phoned to tell her that they had ordered the helicopter and were going to go up to count the birds. They would capture any that they could with a net, he said.

Cynthia could hardly believe what she was hearing. It was again breeding season, and although the population of 'Alala at McCandless had grown from nine to eleven during the past ten years, they were the only wild ones left. "I said, why don't you take your helicopter and fly it over Olinda and see how those birds like it first before you start coming up here and doing this?"

Irate, she called the chairman of the State Department of Land and Natural Resources in Honolulu and convinced him that the maneuver would spell the end of the 'Alala. More headlines—and the state backed down, to the dismay of the National Audubon Society, whose Hawaii representative was quoted in the next day's paper saying they were giving up the last chance to save the Hawaiian crow (Ten-Bruggencate 1989b). This was just one of many times that the story of McCandless Ranch and its endangered crows appeared in the local newspapers, often on the front page.

The National Audubon Society threatened to bring suit against McCandless and all its owners, as well as the U.S. Fish and Wildlife Service (USFWS), charging that, under the Endangered Species Act, the USFWS had the authority to enter McCandless property to protect the crow, and McCandless was required to allow them to do so.

But first the new director of the USFWS in the Pacific Basin asked to have the National Academy of Science examine the situation and issue a report, recommending what action should be taken.

"I was very hesitant about it," Cynthia admits. "I didn't feel I had any choice, but I was very hesitant, mostly because it was literally me against the biologists, and here were some more biologists coming to study the situation."

When the National Research Council panel visited Hawaii and asked to see the wild 'Alala in their habitat, however, Cynthia guided them through McCandless, up the mountain. "We took them up for an overnight, the ones that came, and I was totally impressed. It was just like a lifting of a huge burden off me." She recognized that each panel member was an expert in a field critical to the 'Alala's situation. "For the first time I felt that some competent people were looking into the 'Alala situation," she says. She made up her mind to follow whatever recommendations the panel made: "I couldn't be a one-person stand here anymore."

The Committee on the Scientific Bases for the Preservation of the Hawaiian Crow consisted of Chairman W. Donald Duckworth, director of the Bishop Museum in Honolulu; Tom J. Cade, the Peregrine Fund, Boise, Idaho; Hampton L. Carson, University of Hawaii; Scott Derrickson, National Zoological Park, Front Royal, Virginia; John Fitzpatrick, Archbold Biological Station, Lake Placid, Florida; Frances C. James, Florida State University, Tallahassee, Florida, with special advisers Cynthia Kuehler, Zoological Society of San Diego and Stuart Pimm, University of Tennessee, Knoxville, Tennessee.

Just as Cynthia was beginning to feel relieved, however, the Audubon Society followed through on its threat. It brought suit through the Sierra Club Legal Defense Fund, naming McCandless Ranch and the USFWS as defendants in a suit alleging that they had violated the U.S. Endangered Species Act. Although the National Research Council report had not come out when they went to court in Honolulu Federal Court, Cynthia says the judge put a "huge amount of pressure on to settle it."

At that time, she says, the defendants felt they could not push their position any further. "We could not find anybody to support our position, anywhere, anywhere," she says, recalling how bleak the prospect looked. They asked for time, a week, until the report would be issued, but their request was denied by the National Audubon Society. Only through numerous legal delays did they manage to postpone the decision until the National Research Council report was published. Then, says Cynthia, "I felt our position was vindicated."

The first recommendation in the 136-page report was that a viable wild population be maintained. The committee endorsed an active management program as the only way to prevent the wild population from becoming extinct. "We conclude that there is a high probability that the current population in the vicinity of the McCandless Ranch will become extinct as a result of chance events in one to two decades unless its numbers and geographic range are increased," they wrote.

They *did not* recommend that all the 'Alala should be brought into captivity. "To avoid extinction, the sizes of both the wild and captive populations must be increased to provide demographic and genetic security," they wrote. They recommended joint management of the wild and captive populations as a single unit and suggested that a new "recovery team" be set up, with biologists who were experts in avian ecology, captive propagation, reintroduction, long-term avian population biology, and population genetics, as well as an avian veterinary pathologist and an aviculturist. In addition, they urged that a representative of the private sector, preferably a landowner or land manager, be on the team.

The new recovery team, which was appointed after the report was published, included two landowners, the ranch manager from McCandless as well as a representative from Audubon. As the team met to discuss the crows' habitat in a closed session in 1993, however, Cynthia protested loudly that the proceedings were not public. As a result, subsequent meetings were open to all.

The report had stated that the genetic considerations were minimal. The DNA analysis showed the captive flock to be moderately to highly inbred, but the 'Alala on McCandless Ranch were so close to the wild sources of some of the captive birds that "the addition of wild adult birds from the ranch to the captive population for genetic reasons alone would be expected to provide no more than a very minor measure of new genetic variation to the captive population, and their removal from the wild would adversely impact on the extant wild population's potential productivity."

The report from the experts went on: "Any potential beneficial genetic augmentation of the captive flock should be accomplished by removing eggs from the wild and hatching them in captivity." The wild eggs should receive some natural incubation (five to seven days) before they were removed.

In addition, the report strongly recommended that the state and the USFWS establish at least one major forest preserve along the Kona Coast, where cattle ranches provide critical habitat for native flora and fauna, and that wild 'Alala predators be controlled. (A trapper had been employed at McCandless by Cynthia and the USFWS for several years to control the mongooses, rats, and cats that prey on 'Alala nests.)

The settlement provided in the lawsuit, in the spring of 1992, was modeled on the recommendations of the National Academy of Science report, which Cynthia welcomed and has adhered to. A three-year agreement was drawn up. During the first year, only observation, and nothing more, was allowed. In the second and third years, first-clutch eggs could be taken, but any further clutches had to be left for the pairs to raise themselves in the wild.

During the first year of the agreement, one chick fledged in the wild, bringing the population to twelve. Five chicks were raised from McCandless eggs at a Kona facility run by the Peregrine Fund under contract from the USFWS and were released back onto the ranch in 1993. Three of them were tracked; two disappeared. In 1994, seven more birds were released, and one disappeared. A census in late 1994, however, showed that the total number of birds in the wild had dropped back to twelve. It is unclear what happened to the five 'Alala that disappeared, of which only one was a newly reintroduced bird.

The eggs that were removed from the wild during the second year of the agreement were taken to the Kona facility, where their hatching was overseen by specialists sent from the San Diego Zoo. The National Research Council report, which criticized the husbandry and management of the captive 'Alala at the Olinda site, contained a long list of the changes needed. Though Fern Duvall remained on the staff, a new administrator, Peter Shannon, was put in charge. Hatching rates at Olinda increased, and four fledglings survived out of the thirteen eggs laid in 1994.

The original three-year agreement that settled the lawsuit was followed by a two-year written agreement between the USFWS and McCandless, which followed the conditions of the previous agreement fairly closely. Audubon was not involved in the second agreement.

All of the captive-raised birds are now banded, as the report recommended. They also wear what Cynthia refers to as "little backpacks" containing radio transmitters so they can be tracked by the USFWS. Cynthia estimates that the federal government spends at least two million dollars a year on the 'Alala project. The Peregrine Fund, under contract to the USFWS, hatches the eggs taken from the wild and returns them as fledglings. The Peregrine Fund also has taken over the management of Olinda and is in charge of a new facility near the volcano on the Big Island, whose construction was funded by the USFWS. Cynthia is amazed at the money expended per bird: the last egg taken from the ranch was flown by helicopter to the new multispecies captive-breeding facility near the volcano.

Cynthia enthusiastically endorses the Peregrine Fund, which she says has more than proved its competence. She praises Peter Harrity, who is in charge of the aviary, for his attention to the birds' needs and for his ability to identify each individual by sight and by call. "No one else on the project is able to do that," Cynthia says. All of Peregrine's other employees on the Hawaii project are from the San Diego Zoo, and all are aviculturists.

Peregrine has also shown sensitivity to the wishes of the private landowners. If the landowner has any hesitancy about a Peregrine proposal, the fund seeks a more acceptable alternative. "You don't find that with Fish and Wildlife." Indeed, she questions the level of expertise and education of the people who work with the USFWS, particularly the volunteers who come and go every three months and who provide the necessary staffing to cover the crows seven days a week. Volunteers should at least have a college degree, she believes.

In 1995, no eggs could be found on the ranch; in 1996 only one was found and removed to the Peregrine facility. The radio transmitters have led researchers to two bodies of birds released in 1994, probably killed by the *I'o*, the Hawaiian Hawk (also an endangered species). In all, seven released birds and seven of the original wild birds have died, leaving twelve birds on McCandless at the beginning of 1997. Four more were scheduled to be released. Twelve birds, notes Cynthia, "is where we were in 1992, only then they were all wild."

None of the captive-bred birds have reproduced. Like many other things about the 'Alala, the age at which they reach maturity is not known.

"We have no track record," Cynthia says. "We have a great unknown. How are these birds going to be able to function? How are these birds going to be able to raise their own?" She is encouraged, however, that more is being learned about the crows' behavior. In the first year of the agreement, when only observation was permitted, it was learned that a chick was still was being fed by its parents after one year. "It wasn't until the parents started nesting again that they literally kicked it away—so something happens during that year that's being missed out on when you're feeding a bird with a puppet and letting it out on its own after it's two months old."

Captive-reared crows at Olinda now are also laying eggs, which is additional ground for Cynthia's optimism. She notes, however, that she is the only private landowner on the island who is interested in having the crows released on her property, although other sites have been sought by Peregrine. "They're coming face-on with private landowners' not wanting them to be anywhere around them: no, thank you, we don't want you." Not surprisingly, after watching her struggle, other landowners are wary. One Big Island rancher told us that the last thing he ever wanted was to have an endangered species found on his property.

The federal government has been negotiating to buy property adjoining McCandless Ranch, to extend a nature preserve on the island. Cynthia admits that she has "a lot of mixed feelings" about her potential new neighbors. She continues to be skeptical about the USFWS because it has steadfastly refused to admit that human intervention might be a part of the reason for the 'Alala's near-demise. She also continues to think that the bottom line should be the recovery of the 'Alala; she has little tolerance with "research-for-research' sake." A long-term study on mosquitoes, she points out, has run as long as the study on the 'Alala on McCandless. "They [biologists] know that we have the highest incidence of malaria-bearing mosquitoes that they've found, at all elevations. They also know that 100 percent of the birds that they've released have contracted malaria. So, to me, it's obvious that mosquitoes are a problem, and at the point you know they're a problem, you start dealing with the problem. You don't keep researching. I just find that's a waste of money."

She blames the system: "In order to fund the program, they have to justify their funding and the only way they can justify the funding is by setting up some sort of a research plan." It's a vicious—and expensive—cycle for the biologists, she believes, and she is glad that it is finally leading to some action.

The action—to which she has given her blessing—will consist of taking out or draining the core of downed tree ferns, which act as reservoirs for stagnant rainwater in the ranch forest. They are thought to be the breeding ground for mosquitoes. It is known that all the McCandless 'Alala have had avian malaria: the Peregrine team took blood samples from the crows, treated them when they discovered malaria, and then, after the birds had recovered for a while in the aviary, returned them to the wild. Cynthia acknowledges that without this "human interference" the crows might all have died.

After years of concern and disappointment with so much of the scientific research that has been done on the ranch, she has thought of a sure way to limit it. "I would like to lay down a law," she says, "that nothing that goes on up here is going to be published. That would take care of a lot of things." It is an intriguing proposal but not one that appeals to researchers.

Cynthia, however, has thought it through carefully: "Everything that's published is another feather in their cap, and at some point or another publishing becomes self-serving. I think more emphasis is placed on trying to get the information for publishing than on helping the 'Alala. I think the focus changes."

In the meantime, she keeps the key to the ranch gate, gives all workers a speech in which she lays down the ground rules for working with the 'Alala, and requires that a USFWS supervisor be always present. She has forbidden the researchers from working on other endangered species on the ranch and insists that they restrict their focus to the 'Alala. She is resigned to the fact that sometimes during breeding season "a pile of people are up there." They includes three or more volunteers, the USFWS 'Alala researchers, and sometimes a three-person mosquito crew. Cynthia allows only one trusted USFWS supervisor to climb trees and collect eggs from the nests during breeding season.

Once a month, a "partnership" meeting held at McCandless Ranch is attended by landowners and representatives from the Peregrine Fund and the USFWS. At those meetings, Cynthia grills the team about the current status of the 'Alala and the team's activities. It is not hard to imagine her challenging such research as counting the number of berries on a bush on which the 'Alala feed, if she can't see a direct benefit for the crows.

The most recent development at McCandless has been its tentative entry into ecotourism. Realizing how precious her land is and how limited she is in what she can do with it and still protect the endemic flora and fauna, she decided to run nature tours for people. "It's a shift," she admits, and adds that she made the decision in favor of ecotours after deciding against expanding the cattle business.

Although it is not advertised as an 'Alala tour, the crow is the bird nearly everyone wants to see. So far, no one's been disappointed. "We do get people who have their life lists," she says. "But we really would like it to be more of a nature tour, because who knows how long the 'Alala will be around?" She jokes that during the second year of the ecotours, they had a 100 percent increase: "It went from one to two." She expects to have three "big" tours a year, for twelve to twenty-four people each. She tightly controls the daylong tour to keep the group away from breeding pairs. Leaving the ranch headquarters as early as 4:30 A.M., they travel by four-wheel drive vehicle to about the six-thousand-foot level. They customize the tour to meet the wishes of the participants. Asked if curious 'Alala fly down to check out the visitors, Cynthia replies, "They aren't seeking us out anymore—their days of naïveté towards humans are gone."

Cynthia has had plenty of time to consider various aspects of her battle to protect the 'Alala. She has thought considerably about the issue of rights. "As soon as the suit was filed," she says, "it no longer was a bird question; it became a property rights question." Who has the right to keep whom off private land? Cynthia is convinced that, in light of more recent U.S. Supreme Court decisions, if she went to court today, she would win hands down.

She is keenly aware that the crows remain in a precarious situation. If they do go extinct, she says she would be deeply saddened but would at least feel that she had given everything she could to prevent it. "I guess I would feel that leaving them alone seemed to have the best effect for that period of twelve years," she says.

On a broader scale, she reflects, it is a question of values. "Our family have been good stewards of our land," she says. "I don't feel that there has been anything that I have overtly done to have in any way abetted the demise of the 'Alala."

At the 1996 Hawaii Conservation Conference, Cynthia stood in front of a large audience in Honolulu and told them how that good stewardship was working against the ranch. "We are no longer able to conserve our land in the manner which we feel is the most appropriate, after eighty years of experience and success, without interference from all sides, from people who are intellectual 'experts' rather than experiential," she said. She announced that her application to a state of Hawaii forest stewardship program had just been turned down because "our land was too pristine, there was too much native or natural forest for us to qualify."

She related how, at one of the recent "partnership" meetings, she had shocked the USFWS representative by saying that at times she wished her grandfather had just chopped down every tree on the ranch: if the family had not taken such good care of their forest, no one would be bothering them now. Instead, though, their thank-you had been a lawsuit by Audubon.

Then she asked the assembled group what they thought the financial advantage or benefit from her conservation activities was. The answer, she told them, was none. There are no incentives for making conservation a higher priority, she said, only disincentives.

She listed some: Real property tax rates for conservation-zoned land are higher than those for land zoned for agriculture. The taxes on agricultural land in forest are higher than on pasture. Reforestation is considered a tree crop, which carries higher taxes still. And when, she pointed out, because of financial distress land has to be sold after a family has cared for it for generations, family pride of ownership, sense of responsibility, and love and knowledge of the land are all lost. "The private landowner needs to generate income in order to conserve, and the environmental community needs to understand that this is one of the most important facets of conservation on private land," she said in closing. "We live with a hand-to-mouth budget, we scrape dollars from wherever we can find them and put them into that black hole that we love and we call the Ranch—and with every government regulation, with every superfluous lawsuit, with each demand, that hole gets deeper and blacker, until the time comes when we can't hold on any longer and we will be forced to sell, too." She pleaded for incentives. "The carrot, not the stick; help, not threats; encouragement, not impediments; examples, not decrees; cooperation, not war."

Pan troglodytes. Drawing of a chimpanzee by Lorene Simms.

4

DARWIN, ULYSSE, ROUSSEAU—AND BRUTUS, TOO?

We are losing something extremely precious, a knowledge, something which is destroyed forever. We sort of
have our great-grandmother's silverware which remains.
We don't have anything left of the intact world in which
these animals lived.—HEDWIGE BOESCH

It's a kind of necrology; such a huge number of
individuals are simply not there anymore.
—CHRISTOPHE BOESCH

CHRISTOPHE AND BRUTUS are about the same age
and, although they have been educated in quite different schools of ecology, each has achieved the respect of his peers. Most important, they share a
deep understanding of the challenges and beauty of
living in West Africa's largest pristine rain forest,
Taï National Park in the Ivory Coast. They watched for years as scientists came to this 1,350-square-mile (3,500-square-kilometer) forest
to study a group of chimpanzees that now, even by conservative estimates, balances on the edge of extinction. With only three adult
males left in the group in the spring of 1997, Christophe knows the
situation is precarious. Brutus may know it too, for he is one of the
three.

Christophe Boesch talks quietly in his office at the University of
Basel, Switzerland, overlooking the Rhine. An assistant professor of
zoology, he teaches and writes here, but it is in Africa that he does his
research; he is preparing for another trip to Taï. He is "realistic," he
says, emphasizing that the chimps' future is uncertain. But the chimps

are more resilient than he had once thought, and he sees grounds for hope, at least in the short term. Since he began studying them, this group of chimps has been hard hit: logging has damaged their habitat, poachers kill and sell them as "bushmeat," leopards prey on them, refugees and soldiers from the Liberian war hunt them, and most recently, the Ebola virus has decimated them. Christophe balances his optimism, based on his desire to see this group survive, against a pessimism grounded in his knowledge of all the threats to its existence. But with one more male becoming adult, no recent Ebola deaths, and all the females in the Taï group caring for babies, he is encouraged. There is a chance that the numbers can be built up, that a stable population will endure.

Christophe and his wife, Hedwige, have studied these chimpanzees (*Pan troglodytes verus*) for eighteen years, with continuous support from the Swiss National Science Foundation, which, they gratefully acknowledge, has been "extremely generous." Watching the chimps has been an emotional roller coaster ride, enough to give Christophe an ulcer and to make them both fear that they may never again be able to show their children the chimps among which they once lived. They've observed, in succession, stability within the group, followed by decline, then recovery, high mortality, and again some improvement. As the number of chimps has fallen from eighty to thirty, the Boesches have noted interesting behavioral changes. Recently the chimps have hunted much less during the fall, their former hunting season, probably because there are too few of them now to form an effective team. They concentrate, instead, on gathering fruits, nuts, insects, and whatever else they can find. Individuals have also been observed lately to leave the group for extended periods of time. Now one reason for these wanderings is becoming clearer, at least as regards the females: analysis of chimpanzee DNA (deoxyribonucleic acid) has shown that only 45 percent of the offspring born to females in the group have fathers who belong to the group.

Brutus is by far the oldest member of the group, which lives in a twenty-seven-square-kilometer (ten-square-mile) home range in the western part of the park. When asked just how old Brutus is, Christophe smiles and says convincingly: "He was born on the eleventh of August, '51." Those who are listening laugh, for that is Christophe's birthday. He has calculated that he and Brutus are very close to being the same age. Macho, a male who is almost thirty, was also present when Christophe and Hedwige began their study, as were three old females.

The decline in the number of chimps reflects a general decrease in the park's mammal population. A recent survey by national park officials showed that over the previous eight years the number of mammals in some areas of the Taï National Park declined by more than 80 percent, largely because of poaching. Christophe believes that illegal hunting has the potential to empty a forest. "That's the way it may go in Taï National Park if nothing is done to control specifically for poaching," he says. "It's not a question of not being aware, it's not a question of not being willing to do something, but it's very difficult to control, and actually they should tackle the problem of bushmeat within the country—it's much more difficult to catch a poacher in a pristine forest than to control the bushmeat traffic on the roads." Chimps, which are difficult to find, in fact fare better than many animals that are sold for food; poachers concentrate on easy prey such as forest antelopes and small monkeys.

Christophe and Hedwige have witnessed some major behavioral changes within a relatively short time. In a 1989 paper, which drew considerable attention, the two described how the Taï chimpanzees carried out cooperative hunting expeditions on which they attacked and ate primarily red colobus monkeys. Those expeditions have now become far less frequent. Jane Goodall had earlier documented hunting, mainly by solitary chimps, but social hunting with the frequency and intensity that Christophe and Hedwige observed had never been reported. They have also shown that tool use among Taï chimps is more highly developed than that of any nonhuman primate. They use hammers and anvils to crack nuts, a technique they can teach their offspring. The closeness of chimpanzees to humans has fascinated Christophe and Hedwige; their observations have raised questions about the history and uniqueness of humans. A better knowledge of chimpanzees, observes Christophe, may change the way we see ourselves.

It has been known since 1976 that chimpanzees, which share 98.4 percent of their DNA with humans, rank among our closest living relatives. (The pygmy chimpanzee, or bonobo, may be still more closely related.) The Boesches have been credited with showing that the boundary between animal and human nature is less clear than might be expected; the similarity between human beings and chimpanzees is revealed in a variety of ways. When chimps crack nuts, they actively shape the hammers to the task. They keep a mental map of the hammers and the nut-cracking sites in the forest that enables them to choose the nearest hammer of high quality, even when the anvil is out of sight—an ability the Boesches compare to that of a nine-year-old human. Females crack nuts with their offspring, share the nuts with them, and as the offspring grow and start trying to crack nuts themselves, the mothers correct their technique.

These discoveries, and the Boesches' closeness to the silverback Brutus, came after years of patience and close scientific observation. From 1979 to 1991, Christophe and Hedwige lived in a thatched-roof house in a clearing in the rain forest and followed and documented every move of Brutus's group. They worked with Ivorians, searching every day for the animals, which, in the beginning, they heard but seldom saw. In a 1981 paper they reported that forty-two hundred field-hours had—until then—yielded only sixty-two hours of actual observation of the chimps. Christophe still identifies the chimps by parts of their bodies—an ear, a part of a back, the rump—because for years the Boesches were rarely able to see a whole individual in the forest.

The careful observation was mutual; the chimps were taking a good look at them, as well, and eventually allowed them to get close. As the chimps became habituated to their presence, Christophe and Hedwige were learning to recognize individuals. Habituation took nearly three years of going into the forest before dawn nearly every day in hopes of catching a glimpse of the apes, and of following them as closely as possible without disturbing them, before trudging back to camp to write out the observations.

The Boesches' patience paid off; they have published many often-cited scientific articles and are mining their years' worth of data for material for further articles and books. The irony is that their subjects, now well known to the world's primatologists through publications and to the general public through television documentaries, are disappearing from the forest. At thirty members, the group now comprises about the minimum number necessary to survive.

For Christophe it has meant the fulfillment of a childhood dream to observe these apes in their natural habitat; he still takes genuine pleasure in being alone with them in the forest. He feels fortunate to have seen chimpanzees, gorillas, and orangutans in the wild and to have studied two of those three species.

He did a master's degree with Dian Fossey in 1973, on mountain gorillas in Rwanda, before returning to Switzerland to complete a Ph.D. in Zurich. Originally, he had wanted to study gorillas, but on recognizing the complexity of the political problems in Rwanda, he turned to the chimpanzees of the Ivory Coast. The French mammalogist François Bourlière had told him that he might find a nut-cracking culture there. He has also studied Jane Goodall's chimps at the Gombe Stream National Park in Tanzania and compared them with the Taï chimps. Scientific research on great apes throughout Africa is beset with nearly overwhelming difficulties and frustrations on a continent plagued by political, social, and economic turmoil. Fossey, for example, was killed at her mountain camp in Rwanda in 1985.

Through the years, Christophe's affection for Brutus has grown. The silverback is not only the best hunter in the group but has taken a mentor's interest in young males and has groomed several to become good hunters. When the mother of a chimp named Ali died, for example, Brutus adopted him, and Ali began hunting at a much younger age than is common. Brutus, a crafty defender of the group's territory, has conducted successful raids on larger groups, while avoiding potentially suicidal confrontations.

In recent years, though, Brutus has displayed some uncharacteristic behavior. He and one of his consorts, Héra, disappeared from the group for twenty days in 1994, Brutus's longest absence on record. Worried that the two had been killed by poachers, a hungry leopard, or perhaps a neighboring group of chimpanzees, Christophe and his longtime Ivorian aide, Grégoire, kept an anxious watch. When the chimps reappeared, together and uninjured, Christophe breathed a sigh of relief that Brutus was still alive—perhaps he had merely been enjoying a second honeymoon. In the spring of 1997 word came again from Taï that Brutus had disappeared. Christophe, obviously anxious about the report, was eager to get back to the forest to learn whether there was any sign of his old friend. The loss of even one adult male at this point could spell the dispersion of the whole group, for the adult males are needed to protect it cooperatively from attacks by leopards and neighboring bands of chimpanzees.

"I have reached a state where I have more or less accepted that they are probably going to disappear," Christophe explains, and tries to assess his reactions objectively. "I have the scientific one, which, put to an extreme, is to say we have the pleasure to be able to document the extinction of a natural group." He, like other scientists, knows that extinctions have occurred—and continue to occur—frequently, with many species, but usually no one is around to record them.

He acknowledges that his other reaction to their extinction, however, is simple and emotional: "It's something awful." Chimps, he fervently believes, should like other animal species have the right to live on the planet, and it is up to humans to make this possible.

Christophe and Hedwige's stand on this issue is not surprising; they have had more occasion to think about it than most. They have lived in the rain forest, raised their own two small children in the Boesch *campement* (encampment) and returned to Switzerland in 1991 only to put the children in school and to improve their contact with the academic community. Now, as they analyze their work, Christophe and Hedwige react with resignation to the fact that three-quarters of the names in the data belonged to chimps who are now dead. "It's a kind of necrology," Christophe says grimly; "such a huge number of individuals are simply not there anymore."

Is it true, asks a colleague who visited Taï in 1989, that Darwin, Ulysse, and Rousseau are all dead now?

Yes, Christophe nods slowly, on hearing the names that he and Hedwige gave to some of their chimps. Those three represented half the adult males at the time; they disappeared within five years. Eight more chimps were lost in 1992 and twelve in 1994, all from disease. Only one sample was obtained from those twenty: the cause of death was confirmed as Ebola.

The story behind the sample actually grabbed more attention than much of the laborious scientific work on the chimps' behavior. It is an interesting chronology:

In November 1994, one of Christophe's graduate students, a young Swiss woman, took a sample from a chimp found dead in the forest. After putting the sample into a jar and screwing on the lid, she removed her shoulder-length gloves to write the label for the blood-smeared bottle. About a week later, she fell ill and was hospitalized in Abidjan with acute fever, diarrhea, and a rash, then evacuated to Switzerland, where she was treated in Basel's Kantonsspital, the city's main hospital. To everyone's great relief, she recovered fully. Before returning to Africa, however, she was told that medical experts at the Institute Pasteur in Paris and the Swiss Tropical Institute in Basel had determined that her illness had been neither malaria nor dengue fever, as first suspected, but a new type of filovirus closely related to the virulent Ebola virus that had killed many people in Zaire and Sudan in 1976 and 1979.

This information was disturbing to all who were involved in the Taï chimpanzee project, but in an effort to stave off panic and protect the young woman from unwanted publicity, they decided that no public announcement should be made until further details were known. It was a surprise, therefore, when on April 11, 1995, a startling headline was splashed across the front page of Switzerland's largest tabloid, *Blick*: "Death Virus! Ape Infects Swiss Researcher." The article reported that a researcher studying chimps in the Ivory Coast had contracted Ebola and then recovered in a Swiss hospital. The sensational headline coincided with the opening of the movie *Outbreak* in Basel, as well as with the success of Richard Preston's book *The Hot Zone,* then on the international best-seller list. Both brought notoriety to the gruesome symptoms and devastating consequences of Ebola.

The Zoology Institute in Basel was besieged by journalists and television crews wanting to get in touch with Christophe and Hedwige and the unidentified graduate student. Christophe and Hedwige were away at the time; the head of the institute put out a message on the Internet, on Primate Talk, urging restraint and announcing that forthcoming issues of *Science* and the *Lancet* would run articles by Christophe Boesch and scientists from the Institute Pasteur in Paris that would clarify the situation.

On April 26, 1995, in their office in Basel, Christophe and Hedwige unpacked the skeletons of the twelve chimps that had died in Taï during 1994, probably from the Ebola virus. Some would be used to help determine whether the West African chimps actually constitute a separate species of chimpanzees.

On May 6, 1995, Christophe departed for Taï to rendezvous with scientists from the Centers for Disease Control and Prevention (CDC) in Atlanta, who had requested his aid in taking blood samples from the remaining chimps to determine if they were infected with an Ebola virus. They also wanted to interview the Swiss graduate student, who was again working with the chimps, about her infection with and recovery from the new Ebola virus. The goal was to determine the natural reservoir for this Ebola virus; researchers believed that it could not be the chimpanzees, because they seem to die very quickly from it. They planned to test as many plants and animals as possible. Attention was also focused on the fact that it had been during two Novembers, near the end of the rainy season, that Christophe's chimp group had lost a large number of animals to mysterious illness. The question arose: What happens in the rain forest in November that could contribute to an outbreak?

Banner headlines in newspapers around the world on May 11–12, 1995, announced an outbreak of the Ebola virus in Kikwit, Zaire, from which many people had already died. The team from the CDC that had been sent to the Ivory Coast to meet Christophe was immediately diverted to Zaire. On May 18, the road from Kikwit to the Zairian capital of Kinshasa was blocked, as health officials tried to contain the movement of people from the affected area. The World Health Organization (WHO) reported that seventy-nine people had died from Ebola, and that the mortality rate was nearly 70 percent.

On May 19, *Science* published a news article on the chimps at Taï, "Chimpanzee Outbreak Heats Up Search for Ebola Origin," saying that the Taï situation might offer valuable clues to determining the reservoir for Ebola in Africa. "If so," it stated, "the wild chimpanzees that Christophe Boesch had been studying since 1979 in the Taï National Park will have made a major, if tragic, contribution to virology."

Then, on May 20, the medical journal the *Lancet* published an article, "Isolation and Partial Characterization of a New Strain of Ebola Virus," written by Christophe Boesch along with scientists from France and Africa and focusing on the graduate student's illness, with photographs of the new type of Ebola virus, which they described as "serologically related to, but distinct from, the deadly Ebola Zaire." In Basel, the daily newspaper printed a story about Christophe's *Lancet* article alongside articles reporting that the Ebola virus in Kikwit had claimed its 108th victim. The headline read: "Basel Researcher

Enormously Lucky!" By May 30, 1995, 205 cases of Ebola had been confirmed in Zaire, and 153 deaths.

In the end, the CDC team never made it to Taï, but the U.S. Department of Defense, working with French veterinarians in the Ivory Coast, continues to look for the reservoir of the new strain of Ebola in the forest where the chimps live. More than three hundred samples have been taken from rodents, birds, bats, and other animals in their search, but the host has not yet been found. In the summer of 1996, two red colobus monkeys were found dead; a sample from one of them tested positive for Ebola. No more chimps are known to have become victims.

The graduate student whose illness drew headlines has returned to Basel, and, in accordance with her wishes and the Boesches', her name has never been released. Christophe and Hedwige do not allow students or field assistants now to touch any dead animal found in the forest; the WHO is to be informed as soon as possible about any dead animal.

Over the years, as Christophe and Hedwige have watched chimps die, they have been tempted to interfere and help a wounded animal. Only once did they administer antibiotics—to a chimp named Falstaff, after a leopard attack. Otherwise they remained objective, watching, recording the chimps' lives—and deaths.

Hedwige's description of the uniqueness of the rain forest makes clear that her move back to the city was difficult: "The feeling you have inside this forest you have nowhere else, the noise, the different songs, the plants, the mist," she says softly. "You can't find that here." Although Christophe continues to spend at least three months each year at Taï, Hedwige and the children are now limited to a part of western Europe where "wildlife" means a swan on the Rhine or a hedgehog in the garden. Yet, the chimps are still a part of their lives.

Once in Basel they mentioned to their children the annual Darwin's Birthday Party that evolutionary biologists in Switzerland celebrate each February in honor of Charles Darwin. Darwin was also the name the Boesches had given to one of the Taï chimps. When their son, Lukas, heard about Darwin's party, he said he thought it was a lot of fuss for a chimp's birthday party. "Then we told them that Darwin was not only a chimp," Hedwige recollects with a smile. Christophe adds that he hopes there are chimpanzees still to see in Taï when Lukas is old enough to go with him into the forest to follow them.

Hedwige recalls driving from Abidjan to Taï with Christophe for the first time in 1976, through a green tunnel under the forest canopy. "And I thought," she whispers, "that's the end of the world!"

Her excitement dies down, and she describes what is left simply as "a souvenir." Some of the forest will be preserved, she believes, because farmers are beginning to understand that they need the forest: "They have learned that forest means rain in a way: the desert is where there is no forest and there is no rain." She is less sanguine, however, about the chances of the animals who live in the forest.

"We are losing something extremely precious," she says, "a knowledge, something which is destroyed forever. We sort of have our great-grandmother's silverware which remains. We don't have anything left of the intact world in which these animals lived. It's just tiny precious things we keep . . . over some more generations and finally [we] lose it somewhere. I don't think there is anything left really intact anymore, so the chimp is just one species more that will be gone."

She recalls hiking as a child with her father in the forest in Switzerland and once picking a wild orchid called a *Frauenschuh* (lady's slipper), to take home to her mother. That provoked a lecture from her father that she has not forgotten. "He said, 'You must never take these orchid flowers: they have to be here, they have to stay here.'"

Their scientific work, which has been exacting and tough, has produced impressive data. The physical stamina required of them has also been formidable. A typical day begins at 6:00 A.M., before the sun has risen. Breakfast in the thatch-roofed hut of the Boesch *campement* consists of hot tea, fruit, and bread. Christophe and Hedwige, having stuffed oranges into their pockets for lunch, enter the forest by 6:30. Having left the chimps about an hour to the northeast the previous night, the two move quickly to catch up with the animals before they are too far away. Even without the benefit of paths in the forest, the pace is just short of a trot.

The forest is dark. Overhead, three to five layers of foliage lie between them and the sun. Even with special high-speed film, photography is difficult. Visibility is never more than about twenty meters.

The understory, however, is fairly open: it is too dark near the ground for many small trees and bushes to grow. In the occasional clearings where trees have fallen, the vegetation forms real obstacles. From time to time they are caught in the golden nets of large spiders, whose thick silk anchor cables seem to have been designed to catch birds; it is an effort for a man to break through. Near the clearings, thick vines with long thorns can convince even elephants to make a detour. In the deep forest the understory is thin and the trunks of the canopy trees, clear to a considerable height, give the impression of the columns of an open-ended cathedral, extending for kilometers in every direction.

The Boesches pause every five minutes or so to check the compass direction and to listen. Even after years in the forest, they always carry compasses—they are too far under the canopy to tell the direction of the sun, and the topography is gently rolling, similar in all directions. (Ignoring warnings not to go out without compasses, some students visiting the Boesch *campement* entered the forest before dinner one evening for a short walk. They emerged three days later, tired, scared, and very hungry.)

Christophe and Hedwige listen carefully whenever they pause. Chimps are noisy, it should be possible to hear them before long. It has been an hour and a half since the Boesches left the hut. If the chimpanzees are in the same shallow valley, they can be heard up to half a kilometer away, sometimes more, but if even a small ridge comes between, they cannot be heard more than a hundred meters away. Christophe cocks his ear and he and Hedwige look at each other: off to the right, drumming—the chimps' telegraphic communication. Suddenly, black shapes appear, first to the right, then to the left: female chimps, walking purposefully.

Once contact is made, the humans quietly take out their notebooks and pens and watch. The chimps know they are being watched; they tolerate the intrusion because they have "known" the two for years.

This type of quiet observation is entirely different from going on a hunt with the male chimps—a hair-raising experience, like being point soldier on patrol in enemy territory. The males move swiftly through the forest with silent and deadly purpose. It is hard to keep up: chimps are in good shape, humans less so. While some chimps climb into the canopy to chase or herd monkeys into a trap, others approach from the ground or from another direction through the trees. The capture is brutal, the feeding violent and gruesome. Often the victim does not die for several minutes; the lucky ones are dispatched immediately with a neck bite. When a capture is made, the forest resounds with a cacophony of cries from excited chimps, male and female, young and old, rushing in a frenzy to get a piece of the meat. Even for scientists who have observed it often, it can be frightening.

While hunting by the Taï chimps may be diminishing, the chimps themselves continue to fall prey to leopards. When food is abundant and leopards are scarce, the chimpanzees form large mixed-sex groups, but when food is scarce and leopards are abundant, the groups are smaller and the adult males stick together. If a leopard attacks a chimp, the adult males rush to drive it off, screaming wildly. Still, in any given year, almost one chimp in three will be attacked by a leopard and one in twenty will die as a result.

Poaching by humans is an even greater cause of mortality. Usually, Christophe says, the bodies are never found, so the amount of poaching cannot be quantified. It appears to be increasing, however; both Ivorians and Liberians have recently been arrested within the park. The village of Taï, on a main north-south road, lies about twenty kilometers (twelve miles) west of the park and only a few kilometers from the border with Liberia. During the war in Liberia, refugees as well as soldiers crossed the border into Taï and also into the park, to seek safety and food. Chimps presented a sizable target.

The human population around Taï more than doubled between 1992 and 1994. Refugees were housed with Ivorian families, some of which already had found it difficult to feed their own. Since the same tribes live on both sides of the border between the Ivory Coast and Liberia, it seemed natural to the Ivorians to take in their refugee relatives. "The Red Cross has said they've never seen anything like such an integration into the families," Hedwige comments. "But of course, they can't keep them in their own house forever, so they give them patches of land; they make a hut and they plant some rice and they hunt to get food for their children." Only when a refugee camp was opened at Guiglo to the north was some of the pressure taken off the Ivorians, the park, and its animals.

Hunting is synonymous with poaching in the Ivory Coast, for all hunting in the country is illegal, although the law is loosely enforced. Usually two or three poachers supply meat for each village, but with the country's growing population, the demand for meat is increasing. In May 1994, a law was passed by the national assembly that would allow hunting, except in national parks. The law was not signed by the president, however, and has not taken effect.

Christophe is realistic about the need for hunting. The poachers, he acknowledges, "need meat like everybody." He explains that they usually come for the most abundant species, but "if while moving through the forest, they happen to find something else, they will not avoid it." Thus chimps, like everything from rats to antelope, fall prey to poachers. In the fall of 1994, Ivorian authorities arrested eighteen people inside the park near Christophe's group. Two-thirds were gathering fruit by cutting down the fruit trees. One-third were poachers.

Christophe rejects the question, common in conservation circles, about what an animal or plant is "worth." He cautions that this way of thinking is extremely dangerous. "It's not a question of 'worth,'" he says sharply. "If you stop thinking 'worth,' you can start with respect and I think maybe you would save a lot of life on this planet." His estimate of the value of the chimps? "Priceless."

In Africa, Hedwige says, one sometimes hears the chimps described as "half a human." She tells about the director of the Institut d'Ecologie Tropicale in Abidjan, who at first didn't understand the word *chimpanzee*. "Then we spoke more about it and he said, in French, 'Ah, *singe pensé . . . singe pensant*,' which means 'thinking monkey.'"

The Boesches credit the former president of the Ivory Coast, Félix Houphouët-Boigny, with creation of Taï National Park and, to a certain extent, with a policy to protect it from logging and poaching. "He made it clear to the people that the national park was intended to remain a national park and not a poaching reserve, which is certainly not the easiest message to give to the people," Christophe says. When they first began working in the park, logging was still allowed there; soon afterward it was banned.

Houphouët-Boigny, who died in December 1993, was also behind the Ivory Coast's policy of unrestricted immigration, which led to increased population and, paradoxically, to greater poaching pressures on the national parks he wanted to protect.

The Boesches both think that although the forest may survive, its animals will disappear, leaving behind what Christophe sardonically calls "a nice national park on paper." He cites the situation in another national park in the Ivory Coast where students spent several months but saw monkeys only twice, one duiker, and a few birds, "which means, you can have a perfectly protected forest, but empty, simply empty."

Taï National Park still has a chance because it has become a focal point for conservation organizations such as the World Wildlife Fund (WWF). Hedwige remembers hearing people who lived in the Ivory Coast in the 1960s and 1970s tell of seeing wild elephants between Abidjan and the Centre Suisse, in a suburb. "When you travel now from Abidjan to Taï, what do you see?" she asks. "Nothing, no trees, let's not even talk about forest; there is nothing pure left." Taï National Park now is, in her words, "the last bit which exists."

Though farmers in the area long for a piece of the fertile land within the park, Christophe knows giving it to them would be a short-sighted solution. "If we now give the forest to farmers, it will be a solution to their problem for maybe three or five years, and that's all," he says. "Then they will face exactly the same problem but without any pristine forest left, so it's not a solution to the population problem—it may be actually a risk."

He laments the "dollar culture" that has spread in Africa, as cash crops such as cocoa and coffee lure farmers away from their traditional rice fields. Farmers, eager to have motorcycles, clothes, and school tuition for their children, abandon their rice fields to cultivate coffee and cocoa for cash. When they run out of rice, the money is gone, and so is the time to plant more rice. The result is hunger, even starvation, previously unknown, in this fertile region. "The old people remember the good times," says Hedwige sadly, "when there was rice twice a day. Imagine, with all this 'progress'. . . "

The dryland rice grown in the Ivory Coast needs rich soil, like that found under the intact forest. After about three years of producing rice on the same plot of land, farmers normally move on to cultivate a new piece. They now have difficulty finding the rich land for the new rice fields. Rice production is lower on poorer soil, and farmers must therefore make a quicker turnover. In the cycle thus set in place, both nature and farmers lose out.

Rice is of great nutritional importance to the Ivorians, for even with bushmeat, they eat comparatively little meat and consume no dairy products. The local variety of cow does not produce milk; nursing babies are weaned onto rice. Fish is the most readily available source of protein in the diet, and less expensive than meat. The nutritional dilemma is compounded when farmers spend precious money to buy French bread, increasingly available in much of the Ivory Coast. Although foreign to their diet, it has become very popular.

Both Christophe and Hedwige worry about efforts to transfer European and American standards to African people, who, they suggest, have different ideas about the future of Africa than do Europeans. "Many realize now that they did not want what they got," Hedwige says. She relates how artificial lakes were built in Africa with the latest imported technology in order to supply electricity: "They destroyed huge places, and now there is no electricity because the lakes are empty." She points out that in Abidjan, which lies five degrees north of the equator, some buildings are now more than twenty stories high. Without electricity, neither the elevators nor air-conditioning can be used.

Christophe sees another danger: the effect of ethnocentrism on the biodiversity crisis. As a scientist who travels widely, conferring with others concerned about biodiversity and extinctions, he has heard numerous proposals both from the scientific community and from politicians. In Europe, he notes, "sustainable use of wildlife," the idea that natural resources should and can be used without being endangered or destroyed, is increasingly popular.

He tells of a suggestion from a European agency that Africans be able to benefit from Taï park without damaging it. "The idea was actually to give them the opportunity to collect nuts within the park, which would be the kind of renewable product that could be taken from the forest without destroying the forest." The suggestion was designed to help farmers living around the park who cultivate cocoa, rice, or other crops.

Christophe is incredulous. "What they are looking for is not to go back to the forest as gatherers!" he exclaims. "They are for progress, they want to increase the living standard, which would mean a television, a better house or sanitary conditions, hygiene, these kinds of things—but not going back to the forest." When such misunderstandings exist between people of different cultures, it is not surprising that human beings misunderstand other species.

In the spring of 1995, Christophe and his students succeeded in habituating a second, neighboring group of about sixty chimps. That process took five years. Habituation of a third neighboring group, of undetermined size, was well under way by the spring of 1997. In the summer of 1998, Christophe and Hedwige moved from Basel to Leipzig, where Christophe became a director at the new Max Planck Institute for Anthropology. The institute has ambitious plans to build expanded facilities for studying chimps in captivity, but Christophe's work in Taï continues.

The habituation of the second and third chimp groups was not just a hedge against the disappearance of the first group. Christophe is looking at intercommunity interactions and, using DNA analyses, studying the paternity within and between groups. A third part of the studies will focus on the cultural differences between the chimp groups and what kinds of efforts they make to differentiate themselves.

He estimates that the national park as a whole may be home to fifteen hundred chimps, but he is not optimistic about finding any single large community of chimps in the forest. Recalling that Brutus's group numbered more than eighty when they began studying them, he says it is clear that although there is no social limit on size, that other factors determine it. He knows that Ebola took a toll on both the habituated groups and admits that he is afraid of another outbreak.

Using DNA analyses, scientists can learn even from chimps that are no longer alive. It is a bleak prospect, but "we have the genotype and we can still work," Christophe says soberly.

"We don't know what the future is," he concludes. "In French we would say, *l'épée de Damoclès*, the sword of Damocles, is hanging over our head, and the chimps could disappear within three months." This has happened in other small populations and he knows it could happen in Taï.

Pascal Gagneux, Christophe's first doctoral student, moved in 1994 from Basel to a laboratory at the University of California, San Diego (UCSD), in La Jolla, where he learned to sequence DNA from the chimpanzee hairs he had collected in Taï. He had spent considerable time in the forest, first as a master's candidate observing mother-offspring relationships, later as a doctoral student using mountaineering techniques to climb on a single rope into the forest canopy to collect chimp hairs from nests in tree tops. This physically challenging feat, for which he had prepared in Basel by climbing ropes and working out with martial arts instructors, is no longer used: researchers have since learned they can collect saliva samples on the ground from wads of vegetation that the chimps spit out after extracting the nourishment.

Pascal also visited other areas in West Africa to collect hair samples so that he could use the DNA sequences from different chimp populations to estimate their relationships. Working with the DNA from Taï, which, he emphasizes is from individuals with known histories, "not out of some gene bag," he is gaining insights into the social lives of chimps which were not clear in the forest.

Employing the technique of polymerase chain reaction (PCR), he has used the small DNA samples available in a few cells at the root of a single hair to characterize eleven nuclear genes (loci) for wild chimpanzees. At such loci, individuals inherit a gene from each parent, and the combination is unique for each individual. This allows identification, paternity testing, and studies of relatedness and subdivision in wild populations. He has also analyzed mitochondrial DNA from populations across West Africa to study gene flow and population differences over broader geographic ranges.

His analyses have shown that only 45 percent of the offspring born among the Taï chimps have fathers in that group, meaning that the majority of the babies are fathered by males in one of the four or five neighboring groups, a much higher percentage than Pascal or Christophe had suspected. It is a discovery that makes Christophe smile ironically: he had always thought that the reason young females frequently transferred to a neighboring group was to avoid inbreeding. Now it appears they may be going back to their parents' home, where they may mate with their fathers or brothers.

When Pascal, Christophe, and their colleague David Woodruff at UCSD published these findings from Taï in May 1997, they refuted the long-held assumption that chimpanzees' social unit and their reproductive unit in communities are identical. The *Nature* article "Furtive Mating in Female Chimpanzees" concludes with the comment: "That mate choice can occur within a larger area than the traditionally observed social units has implications not only for the evolution of mating systems and sociality, but also for the management of viable populations of these threatened primates."

Pascal has an additional reaction to their findings: "The fact that females exert their mate choice across the boundaries of their social communities is esthetically pleasing to me, especially considering how much shit they take from the males in their community."

The lab work has also given him opportunity to think more about how the chimps and other apes are related to humans. He now feels that humans and bonobos, the pygmy chimps living in Central Africa, could be sister species that diverged only after having split from the common chimps. "This feeling is based on the phylogenetic trees we obtain when we combine the human, bonobo and chimp data," he explains. "In the past people would use one or two laboratory chimps of unknown origin as the standard 'chimp,'" he adds. But his work with DNA has shown him a huge variation among chimps. In West Africa, the chimps display more than three times the genetic variation documented for all modern humans in the same region of their mitochondrial DNA.

By the time Pascal returned from Taï to begin his analyses, he had been deeply discouraged by the shrinking numbers in the chimp group and by the pervasive presence of poachers within the park. "The word for 'animal' in the Ivory Coast is *viande*," he explains, using the French word for meat, "and they eat them all." The broad menu of bushmeat that he lists, considered to be "very *doux,* sweet, very good," includes rats, antelope, monkeys, and chimpanzees. He does point out that people differ; there are *totems,* and certain animals that some people do not eat. "So in a certain village you might have two families that do not eat chimpanzees, but all the rest do, and chimpanzee is better than most, considered by some to be very special." Another advantage of chimpanzee, he adds, is that it is large: "With one shot you have forty kilos [eighty-eight pounds] of meat."

He sighs and admits he knows that most of the local assistants who work with biologists in the field would buy bushmeat if they saw it at a local market. "And they make this little excuse for themselves: they say it's already dead, they're not killing it. They don't see their demand as creating the market." He pauses. "If you asked people working with chimps, they would probably say, 'we would not eat chimp meat,' I'm very sure of that. But—they would probably eat colobus." He laughs and shakes his head. To non-Africans, at least, it seems very arbitrary.

Pascal believes that most Africans realize their actions have had an impact on the animals and environment. "Most parents in Africa grew up in villages where you could hear monkeys every night and day," he says. "Most sons and daughters at this stage grow up in places where you never see or hear monkeys. Most parents had troubles keeping elephants out of their fields only one generation ago, and a generation in Africa is so short."

When Pascal left Basel in October, 1993, to crisscross the park and collect samples from nests, he was accompanied by Barbara, his wife of a few months. They launched into the project with great enthusiasm and energy. What they saw affected them deeply. Pascal speaks reflectively and very seriously about his impressions.

"I doubt that there will be an authority strong enough to protect the park," he says, citing the number of people in and around Taï as well as the economic interests of the Ivory Coast and its people. "I don't think the political will exists to protect a site just for its natural values or its wildlife values. The park is totally enclosed by people, and the population is growing all around it. There's a war on one side and a massive economic crisis in the country, and that's a new problem." He is referring to the war in Liberia and to the devaluation of currency in the Ivory Coast by 50 percent in January 1994, which was preceded by the collapse of the world cocoa market. (The Ivory Coast is the world's largest producer of cocoa.)

Like Christophe and Hedwige, he has observed that the density of wild animals within the park rises as one nears the Boesch camp. "Near these scientists' camps, you hear many more monkeys, you see more tracks, you see duikers," he says. Then, his voice falling, he adds, "And you still see some signs of poaching, you hear gunshots near Christophe and Hedwige's house."

But he was particularly frightened, as he traversed the sixty-kilometer-wide park, by the big poaching paths and the camps right in the middle. He counted up to thirty-one beds in a camp, beds made with branches so that the poachers could sleep above the ground in the rainy season. Not surprisingly, at those places he saw and heard fewer animals. And he notes that even birds such as hornbills had become prey to poachers. "Once they start poaching birds, it's really bad."

Pascal describes in detail the various kinds of traps he has seen, both in the forest and around the farmers' fields. All fields are surrounded by traps, and if the fields lie in an area where hunting has not been intense, big animals like chimps, antelopes, or leopards are caught. "Chimps usually get out," he says, "although they might lose their hand or foot."

The traps he describes are often exquisitely crafted. Usually they incorporate a spring, often a fresh sapling about the thickness of a man's arm, which is bent and attached to a cable (or, to cut down on the cost, a vine). When an animal puts a foot through the loop, the motion triggers the spring and causes the branch to snap up and close the loop around the appendage. The animal is jerked into the air. "Usually when you come on the animal, it is hanging," Pascal explains. "If it's a small animal, you can pick it like a fruit; if it's big animal, a leopard or a chimp or a big antelope, it's trying to get out of the trap." The bigger the animal, the more slowly it dies.

Pascal is no newcomer to environmental and ecological problems. Active in Greenpeace as an student in Switzerland, he was not confronted until he went to Africa with the stark realities that have led him to a deeper level of social and scientific quandaries.

"What scared me so much the first time I went to Africa was that you realize that by actually leading a very modest life in terms of material and energy consumption, these people threaten animals; . . . just in order to live from one day to the next, they have to destroy incredible quantities of sometimes pristine forest."

In 1990 and 1991, he studied mother-offspring relations among the chimps at Taï, watching as baby chimps were born, groomed, trained, protected, and disciplined by their mothers. Many of those mothers, as well as their babies, have since disappeared.

Pascal discovered African society simultaneously with chimpanzee society. And he feels he learned a great deal from both.

"I was there for one year and I knew a little girl. When I came back she was dead. I was told that she was just malnourished, her parents had a salary but they couldn't support her and now she's gone. And I ask myself, why? But then I see: living and dying seems to be so much more intense, happening so much more in Africa. You really see it happening: you see one birth after the other, this young girl has a baby and there an old one disappears, this boy you used to play with died of meningitis, and this little girl died because she was malnourished.

"Life in Africa is very naked, there's not much cultural buildup around it. And this reminds me very much of chimp life because it couldn't be more naked than in this group of beings who live in this forest, sleep and eat and mate and have babies and die.

"What's the value of an individual organism that can be so much fun, that can be so fascinating, that can be beautiful, that can be terrible? And then it's just *not* anymore?"

He said he thought about this constantly as he drove into African villages "with a big car, you know, and some crazy idea about doing science and trying to find some chimps to measure something about their history."

He finds it hard to cope with the deaths of the Africans he knew and with the loss of the chimps he studied. "That's just the way it is, and I find myself just accepting that about twelve or fifteen chimps that I knew just died in four years. And what can I do about it? It's sad, I don't really know how I cope with it—I think I just put it aside. I do not actually cope with it."

Pascal credits Christophe and Hedwige with setting certain standards for scientists and students in the field, such as not carrying machetes. "Their students do not walk around with a sharp blade, cutting down whatever disturbs them," he says. The no-machete policy was one that Christophe brought from his work with Dian Fossey, who never carried a weapon while she was studying gorillas.

During the six months in the forest in 1993 and 1994, Pascal not only encountered poachers but lived with guides he was sure were also poachers. "Their relationship to the forest is really their only resource," he says. And, with a kind of admiration, he describes how self-reliant these men are in the forest. Mimicking their progress through the forest, he gives an animated demonstration, rapidly slashing out with an invisible machete: "We chop this way through the bush—swish—we chop this down to make a nice little bench, if we want to cook—swish—we cut this to make your fire—if you want to eat, you kill this animal . . . "

He stops suddenly and looks up from the invisible swath he has cut. "In a way, it's very nice because you see a person who knows his forest perfectly and is really able to survive in it. But it's so scary because it is one-way consumption. He doesn't do much for the forest; he's just there and can use up everything: 'You want to see this? You want to eat this? You want to collect this? If you're sick, I will go out and get some leaves for you.' So it kind makes the fairy-tale view of a natural people a bit relative."

Although Pascal can identify more with chimps than with certain other animals because of their facial expressions and nearly human mannerisms, in the end he considers them one more biological species, another form of life, all of which is precious. He pauses, then explains: "They happen to sit on the same phylogenetic branch" as human beings, and viewed from that branch, they may seem to stand out, "but not if you sit at the root of the tree, absolutely not. Every end of the tree is then unique."

After working in California on DNA, he realizes he has started forgetting the faces of the chimps from which he collected it. He admits that he misses them, even if he does not miss the life alone in the forest.

Will he return to Taï in time to see Brutus, Macho, and the other chimps again? He believes he will. But he also believes that most chimpanzee populations in West Africa will disappear very soon. "In a few places like Taï, Sapo in Liberia, or Tenkere in Sierra Leone, we might be able to keep local populations going for a while," he says. "I have this sad view of West African chimps meeting the fate of the European bears. If only there were a way to slow the destruction and hunting in the Congo basin, that would be the place where chimps could survive in larger numbers and where females could keep moving between communities."

POSTSCRIPT

In May 1997, Christophe returned from Taï National Park, resigned to the loss of Brutus: he had been missing from the group for two months and was almost certainly dead. No females were missing, and in fact a new baby had been born in the group. Brutus had been showing his age when Christophe last saw him and, at forty-six, was pushing the limits for any chimp in the wild. From DNA analysis, Christophe and Pascal know that Brutus has left at least one off-spring; further analyses may reveal that he fathered others. Brutus may contribute to scientific research for many years to come.

Maculinea arion. Drawing of a large blue butterfly by Lorene Simms.

5

BRITAIN'S LARGE BLUE: A STAR IS REBORN

It was large, it was beautiful, it was *the* most likely
species to become extinct, not just in Britain—it is
gone from large parts of Europe and right across the
Palearctic.—JEREMY THOMAS

IN A BRITISH MELODRAMA, the scenario might go
something like this:

A young graduate student is lured away from his
thesis by a glamorous star with an expensive and
fascinating lifestyle, which, he soon realizes, is based on deception. When he meets her, her fate already appears sealed. He nevertheless believes she is special and tries desperately to save her,
but he fails tragically. Deeply saddened by her death but un-
daunted, he introduces a relative of hers from Sweden to England,
where she is warmly received. The newcomer carries on in the
style of her dead cousin, but alas, she is able to survive only thanks
to the protection of her admirers. She feasts alone on costly delicacies at the expense of her benefactors and conceals her true iden-
tity almost to the end, when her magnificent beauty is revealed
and she soars into the open. Death comes quickly, though, on a
late summer afternoon in Cornwall; but she has left offspring who
have been promised protection.

This is certainly not the way Jeremy Thomas would describe the
disappearance of the large blue butterfly from Britain and his devoted
efforts to reintroduce it, but the basic points of the script are accurate.
Thomas was fascinated with the magnificent creature, as much because of its strange life cycle and the unknown reasons for its disappearance as because of its dramatic beauty. And, like amateur butter-

fly enthusiasts and lepidopterists for nearly two hundred years, he gladly camped on its doorstep to watch the adult's four or five fleeting days of flight before it died. "It was large, it was beautiful, it was *the* most likely species to become extinct, not just in Britain—it is gone from large parts of Europe and right across the Palearctic," he explains. But more than that, he was drawn to it because nothing in the world was more ecologically specialized than the life cycle of "this lovely creature."

Jeremy had completed his undergraduate degree at Cambridge and was in the third year of doctoral work on insect conservation at the University of Leicester in 1972 when it was decided that one last desperate effort should made to keep the large blue from disappearing from Britain. Although he was working on two "hairstreak" butterflies at the Monkswood Research Station at the time, he was asked by the Institute of Terrestrial Ecology to switch his focus to the large blue. Because he knew the situation was crucial, he decided he could finish his Ph.D. later—or as he puts it, he decided to start his postdoc before he had finished his doc.

"By the time I was asked to work on it, in 1972, it was down to two colonies, one of which had about 20 individuals, and that went within the first weeks I looked at it," he recalls, shaking his head. "I watched that colony go extinct; I had never seen a large blue until then." That left just one population of perhaps 250 adults. As Jeremy puts it, "It was very, very late in the day."

The large blue butterfly, *Maculinea arion*, with its spectacular forty-to-forty-three-millimeter wingspan, had been in steady decline in Britain since its discovery about 1795. In fact, its national extinction was predicted in Britain more than a hundred years ago, when about a hundred populations still subsisted there. By the 1920s, conservation budgets included funds to reclaim the sites where it lived, and the first nature reserve for large blues was set up. During the following four decades, five more reserves were established. By the 1950s and 1960s, the number of populations had dwindled to thirty to thirty-five, but some were estimated to have had as many as thirty-five thousand butterflies. Admirers flocked to its sites in July and August to watch the spectacle as the large blues took flight.

The large blue had been the first myrmecophilous (ant-loving) species to have its life cycle worked out in detail. Painstaking observations established that its eggs were laid only on wild thyme, that the very small caterpillars fed on the thyme flowers for about three weeks before they fell off the plant and were picked up in the jaws of red ants and carried underground, where they were cared for by the ants. When some of these red ant nests were dug up a year later, big caterpillars were found in them. They became the magnificent large blue butterflies seen in late summer in the Polden hills in Somerset, the Cotswolds, Dartmoor, and the coastal valleys of Devon and Cornwall.

Little work had been done on the ecology of the large blue since its life cycle was studied in 1915–16. Conservationists and consulting scientists, who had pushed for the nature reserves for the large blue, were stymied in their efforts to pinpoint the reason for its steady decline, but they had made no effort to re-examine its life cycle. It continued to decline, perishing even on the reserves specially created to protect it. Jeremy describes his entry into the picture as "one last desperate throw of the dice to see if there were any other measures which could be taken to save them."

Jeremy went to the south of England to study them, admittedly "not too much of a hardship—it was in a very attractive part of Britain." Camping near the butterfly sites or living in a rented trailer, he began his study, which has continued, with interruptions, for more than twenty years. His work days ran from sunrise to sunset. He began by watching the butterflies' behavior closely, but he also did experiments, for he took the view that when almost everything that had been tried had failed, some unknown subtle requirement for the large blue's existence must be lacking, or something of fundamental importance in its life cycle had been missed by other observers.

He proceeded slowly, however, realizing he was being watched closely now that he was working with the last of Britain's prized large blues. He kept a low profile, literally as well as figuratively, but the large blue's endangered status was already making the news, first in conservation circles, then as a topic of general concern across Britain.

He followed the survival of many individuals from eggs to adults the next year, noting how many eggs the females were laying, exactly where they were laying them, and what their survival rate was. "I had to first of all find out what was killing the eggs, because, as with all insects, you know, quite of lot of eggs were laid," on average about fifty per female. For the population to maintain itself, "they just need two survivors, one male and one female, for the next generation."

It soon became clear that caterpillar mortality in the ants' nests was extremely high. "For some reason no one thought it was likely to be; they thought that wasn't going to be the problem stage," he says.

Although much of the large blue's ecology had been worked out long before, Jeremy contributed a key discovery. At least five species of red ants forage under wild thyme, and all will adopt caterpillars, yet he found that the caterpillar of the large blue survives in the nest of only one red ant species, *Myrmica sabuleti*. To help establish this fact, he used the "cake crumb" technique to trace the different red ant species to their respective homes. He is quick to point out that he did not invent the technique but that it works well. Jeremy favors a Scotch cake that comes with pink, yellow, and white stripes, which make it possible simply to follow cake crumb colors keyed to the various species of red ant. He located areas with red ants and spent hours lying on the ground, watching the insects moving back and forth to their nests. For some ants at hot, dry sites, a simple hole or two, about the diameter of a drinking straw, was the only external indication that an ant nest lay underground. The crumbs, moved along by the ants, created a trail that led to the hidden nests. Near the ants, *M. sabuleti*, living under thyme that grew in the hottest areas, he put one color of cake crumbs as bait; near ants living in other, cooler areas, he placed cake crumbs of other colors.

"Then you watch—rather tedious—you watch the ants picking them up and slowly taking them in. In fact, if you leave them for thirty minutes or so, you go back, and if you are lucky you see a trail of little pink piles there, and follow the different colored trails." He smiles and admits with some embarrassment that he wasn't sure at first if this were even a sensible thing to do. It led him to determine that *M. sabuleti* was, indeed, the host.

"Everyone had grouped red ants all together, just as red ants, and they do look extremely similar," Jeremy explains. "You have to look at tiny markings on their antennae to tell them apart, although underground their behavior and lifestyle are very different."

The caterpillar, Jeremy points out, was living dangerously by tricking the ants into taking it underground to its nest. "It's wonderful if you can achieve the trick, because you live in a very safe place—

no one else can get into the ants' nest. And the ants are bringing all the food to you and feeding their young, which you eat."

In retrospect, he says, it shouldn't have been a surprise to discover that the large blue caterpillar had evolved into a good mimic of only one species of red ant. "But being a good mimic of one species, it means it's a rather bad mimic of other species," he says. In other words, a large blue caterpillar that was carried into the wrong kind of red ants' nest would often be discovered and killed. "Practically always, in the wild, the ants' nests go through a period of stress, with food shortages during ten months. When that happens, the ants become much more picky and start attacking things that aren't giving them exactly the right signals, and what they attack is, among other things, the large blue caterpillars. And they kill them. So by the end of the year, it is quite exceptional for the large blue to have survived in any ant nest except for just this one species."

He notes that relatives of the large blue in mainland Europe use the same technique. Each of them exploits one species of red ant, and it is a different species of ant for each species of large blue. "So the large blue genus . . . has speciated as a single and different parasite for different ants." Because the species of ants look almost identical, "there had been some debate whether the one which is the host to this species of large blue, whether it was a true species of red ant, because it looked so similar to another one. We now know it is [another species]."

Jeremy compares the large blue caterpillar tactics to that of the cuckoo, which lays an egg in the nest of another bird to fool the other bird into raising its young. "It actually tricks the ant into thinking that it's one of its own grubs that has escaped the nest," he explains, with a note of admiration for the blue's cunning behavior. "We now know that once it leaves the thyme plant, it sits on the ground and it simply waits. It actually puts itself in the sort of places [where] these red ants . . . forage most, within a millimeter or two above the ground." Whereas most ants explore the area just above the vegetation, the red ants go underneath. The blues' caterpillars are there waiting, at exactly the right time of day.

Like about a third of the world's butterflies, the large blue's caterpillar has a honey gland on its abdomen that secretes a sweet liquid that attracts ants. But, explains Jeremy, in contrast to most other species, in the large blue butterfly (along with a few others in the world) the honey gland to attract ants is already developed when the caterpillar is quite small. And it uses the gland to great advantage.

He describes his frustration as he lay on the ground, waiting for a caterpillar to be discovered by the ants, the discovery that launches its adoption into the ant nest. Both ants and caterpillar are tiny, so the drama takes place on a Lilliputian scale. The ants have to be within a millimeter or two to notice the caterpillar, and then they have to tap the gland with their antennae to stimulate the secretion of the honey they seek, which contains amino acids and sugars. "But sooner or later, if the ants are there, they find it, and then they find this gland, and they go absolutely wild," he says. "They are all over it, jumping on it, tapping it. Then they usually go back to their nest and recruit a few other ants who sort of tap it, lick it. Usually this goes on for thirty to forty minutes, sometimes up to four hours." Finally, the caterpillar is left with the ant that first found it, and the critical act occurs: the caterpillar rears up on its hind legs, making itself appear to the ant like one of its own grubs.

"So, the ant, instead of thinking it's licking this source of food out in the wild, suddenly thinks one of its own grubs has escaped, and he really does go wild at this stage," Jeremy explains, becoming rather animated himself. "Normally it leaps up and he grasps it in his jaws and waves it around and goes rushing straight back to his nest and puts it in with his own ant grubs."

The caterpillar settles in cozily and promptly starts feeding on the ant grubs among which it has been placed. It lives for the next 10 or 11 months in the ants' nest, always under the guise of being an ant grub. It grows fat, its neighbors become fewer.

Is there any danger that the caterpillar will eat itself out of a food-stocked home by consuming too many grubs?

"Yes, that does happen," Jeremy allows. "In fact, the red ants that support this particular species of blue have comparatively small nests, and many nests are too small to support even one caterpillar. And if you have several taken into the nest—and I've seen forty taken into the same nest in the wild—they all die of starvation."

The ploy works best if a large number of ant nests are available as adoptive homes, sufficiently spread out so each hosts only one or two large blue caterpillars.

He emphasizes, however, that the caterpillar has important adaptations to its strange lifestyle. "One of them is that when it does eat out the ants' nest, it has an unusual ability to starve that most caterpillars don't have except when they're hibernating. It can fast at any time. And what [can] sometimes—but doesn't always—happen, when it grows to a reasonable size and it eats out its ant nest, is that the ants desert the nest, leaving this fat caterpillar behind. And in the wild, on good sites, there's such a pressure on ants' nest sites that often within a week or two or three another colony buds off like bees. They swarm and go out, the colony goes out carrying their grubs into this vacant nest, takes them all in, where there is a fat—or not very fat, by then—starving caterpillar, which can start again. So it takes an awful lot of ants to produce one butterfly."

He estimates that each large blue butterfly caterpillar eats about three hundred large red ant grubs before it finally forms a chrysalis, just below the surface but still in the ants' nest, in late May or early June. It stops eating but continues to produce secretions through microscopic spores in its chrysalis. Jeremy refers to this as a form of "communication" with its hosts, but says the degree of nourishment the ants actually receive from the secretions is negligible.

The large blue hatches out three weeks later, near the end of June or early July. Individual adults, although they could live for three weeks or so, average a lifespan of only four or five days. "Yes, after all that," Jeremy nods. "We're in trouble if we think of it in human terms." He reflects that this is not unusual for a butterfly because they so frequently fall prey to other animals. It was only on big sites, where thousands used to live, that butterfly admirers could count on seeing more than a few large blues.

"Oh, they were much sought after," he muses. "An entire colony, because they do not all emerge on the same day, does last about four or five weeks, the very big ones. But they were really sought out from the moment they were discovered, long before anything about their life cycle was known, because they were so rare and also considered very beautiful. . . . And so they became one of the absolute target species for butterfly collectors. In Victorian times people used to go to the famous sites and spend three weeks just on those sites, collecting all they could."

Jeremy has visited the impressive collections of thousands of large blues at the British Museum and the Natural History Museum in London, but he himself has collected none. He prefers to capture them on film. He hastens to add, however, that no evidence points to collectors as a significant cause in the decline of the large blue.

Taking into consideration the new information he had gathered on the caterpillars' reliance on *M. sabuleti*, Jeremy looked again at the sites from which the large blue butterflies had disappeared. In about half the cases there was no puzzle: the land had either been plowed up or had experienced other fundamental changes in use. In the other half of the cases, the sites looked exactly as they had at the time the butterfly became extinct.

"The real puzzle with this species was that quite suddenly, over a period of three or four years on a site . . . where [the population] appeared to be abundant, [the blue] suddenly nose-dived and it would go extinct on one site, then on another, then another. For no apparent reasons. And it did this on nature reserves just as much as it did on other land, about five sites. The last sites were tagged as nature reserves of some description, and it went downhill very quickly on them as well."

Jeremy checked that *M. sabuleti* were present, that plenty of wild thyme was available for food, and he examined other factors that might be detrimental. He felt his main role was to be a good detective, to check both the obvious and the not-so-obvious.

"People had all these hypotheses and ideas about why it was declining, and for about forty years the conservation measures were based on what were really educated guesses, all sorts of educated guesses," he says. Because one of those "educated guesses" was that butterfly collectors were a cause, conservationists decided to cordon off a few sites as nature reserves. "One of the best colonies was bought in the late 1920s, and they put a great big fence around it," he relates. "And lo, among all the colonies in that part of the world, that one became extinct in three, four, or five years."

He pauses and smiles ruefully. "We now know that it had been killed by kindness, that the early conservationists had unwittingly excluded the natural herbivores that were creating a certain kind of habitat that this butterfly needs."

The ant host, *M. sabuleti*, it turned out, lives only on warm slopes where the grass or turf barely covers the soil and the sun bakes the earth. Because it doesn't build mounds for its nests as some ants do, it requires such a habitat to keep its subterranean nests warm. If grazing by wild herbivores or sheep, for instance, is halted and the grass and thyme are allowed to grow to as little as three centimeters (just over an inch), the shading cools the soil and the *M. sabuleti* die or move to warmer sites. And, of course, any freeloading large blue caterpillars then die underground along with the ant grubs.

No one had noticed. The sites had appeared, to most, unchanged.

On other sites whose soil was thin, it had become uneconomical to cultivate, and farmers had given up trying. In the 1950s, the rabbits that grazed these areas were exterminated by myxomatosis. So here, too, in the absence of herbivores, the turf grew longer and the large blues disappeared. The change, again, was so subtle that no one noticed.

Jeremy eventually surveyed every site in Britain where the butterfly had been seen. In particular, he studied old sites that had not been obviously destroyed. "They still had acres, hectares, of thyme, and on most of them, an abundance of red ants if you parted the grass and looked underneath it," he says. But, in the end, he found that *M. sabuleti* was either rare or completely gone from those sites.

The novelty, Jeremy says, lay in working with tiny differences, the difference between a centimeter or two of grass. Wild thyme, the caterpillar's only food plant (except in southern Europe where large blues lay their eggs on oregano), actually grows slightly more abundantly in the three- or four-centimeter-high turf, so when observers saw such sites, they thought that the wonderful carpet of thyme must be the perfect habitat. "And actually it wasn't," Jeremy says. "You really want the thyme cut back, so it produces quite a lot of flowers" and so the soil gets very hot underneath. The ants liked warmth; it was too cool in the shade.

He knew then that he did not yet have enough evidence to demonstrate that the absence of *M. sabuleti*—or the lack of grazing —was responsible for the large blue's decline. He knew he couldn't stand up in a scientific meeting and say, "Look, this is it," because the data were not yet statistically significant then. But he also knew that time was working against him. "It's incontrovertible now," he reflects, "but then it simply seemed the lead that had been missing and could explain the disappearance from just about every site where the large blue had suddenly and unexpectedly gone."

He is also the first to admit that the discovery did not solve the problem but rather transferred it to another species.

It took Jeremy a while to convince the people in charge of conservation that a major change was needed at the one remaining site. "And then, it took a year or two to argue through committees, then it took a year or two for the changed action to bind, by which time the population was down to tiny numbers—about, I think it was, sixteen individuals. And at that stage, and in fact, for some time beforehand, it can go extinct for any chance reason."

One "chance reason" was a drought in 1976, the worst in western Europe in more than two hundred years, which put both ant and caterpillar populations under stress. At that point, there were five large blues left. The decision was made to rear these in captivity.

"It went against all the principles, all the things I was trying to do," Jeremy admits in describing the situation in which several agencies and conservation groups were involved. But he complied with the decision and was successful in getting one of the last two females to mate, an especially difficult task with large blues in captivity. "So I did manage to get a number of the caterpillars down into ants' nests very satisfactorily and made sure it was *sabuleti*. And I left it until the next year."

Then, in 1978, Jeremy was taken off the job and sent to the continent to work on other projects. He does not talk much about this move, but it clearly still disturbs him. Yet he is convinced that the move made little difference: he was already resigned to the disappearance of the large blue from Britain.

Twenty butterflies emerged the next year, but with such a small number it was decided—in Jeremy's absence—that releasing them was still too risky and they should again be kept in captivity. None of them mated. The last large blues in Britain died without leaving any progeny.

The large blue was the fourth British butterfly to become extinct in the 250 years during which records have been kept. Fifty-eight other resident species remain in Britain, plus a few migrants, which breed in England and are sighted from time to time.

Jeremy was upset at the way the last large blues died, although he acknowledges that they would have eventually gone extinct in the wild. "If I had been around, I would have said, 'No let them fly about, let them mate naturally, then maybe we'll think about rearing some eggs up, doing it again and then letting them go.'"

He recalls the extensive press coverage at the time: BBC news and ITV carried stories, as did many national newspapers and, of course, wildlife magazines. "Once [the butterfly is] gone, people are able to raise the public concern which might have saved it a few years earlier if we'd had the resources to do the research."

He remembers also that although practically no recriminations were exchanged among the large consortium of conservationists who had been trying to maintain the large blue, a number of people who knew little about the species then appeared on radio or television to comment on what had been done wrong and what was needed. He attributes the reproaches in part to the large blue's glamorous image; it was an extraordinary and beautiful butterfly, and very much in the public eye.

In retrospect, Jeremy says he should not have been depressed. "It was only really afterwards, as we got more biological information on it, that one could really predict how slim the chances had been about our saving it." He believes that if scientists had known about its needs just ten years earlier, the large blue could have been brought back from the edge of extinction. Noting that it continued to survive on the European continent, he wistfully concedes: "It seemed a rather special British one, but there was nothing special about it at all."

There was a strong public call to reintroduce the large blue to Britain. It had already disappeared from some European countries before it did from Britain, and clearly the butterfly was even then vanishing from others. With all the newly gained knowledge about its life cycle and habitat needs, there was good reason to believe that it could be successfully reintroduced.

"I was actually about the last person to cope with it," Jeremy remembers. By that point, he had become recognized as *the* authority on the large blue, the scientist who had identified what was needed and who had the most practical experience with the butterfly. But decision making was by then a group effort by at least twenty people, including representatives of several conservation groups, site wardens, national conservation policy makers, and field biologists. When it came to reintroducing the large blue, Jeremy was "really just about the last to be persuaded that we ought to do it."

Why did he hesitate—what made him reluctant?

By then he had studied the large blue's relatives on the European continent, some of which were more endangered globally than *Maculinea arion,* and even less was known of their ecology. He had begun to put the position of the large blue butterfly in Britain into global perspective.

He had a personal reason, too: "I also knew *I*'d have to do it, and I guess—I *knew*—it had been a very long, weary thing trying to save it." In addition to the scientific challenges of reintroducing the butterfly, Jeremy was aware, the battle would involve politics, considerable publicity, and a public that thought it knew better than the scientists. On the whole, public support for the reintroduction was strong, yet some people were opposed to it—not many, but enough to make aspects of the job unpleasant.

"I just knew what a big task it was going to be, and it has been," he sighs. "But I was persuaded; I didn't hold out long."

Was he disturbed by the thought of tampering with a natural phenomenon, the extinction of the large blue in Britain? After all, he had recognized, perhaps too late, just how overwhelming the odds had been.

"No," he answers unequivocally. "And I say no because a large part of conservation biology by then, including some of my own work on other butterfly species, had shown how artificial, in a way, most of the habitats in Britain and a large part of Europe are. They had been interfered with by man since the prehistoric [era], by farming and forestry and so on, and the wildlife we have is a legacy of the species which were conditioned, created, or destroyed under those conditions." He points out that there are only a few areas, for example high mountain habitats or polar regions in Europe, where this doesn't apply.

"It's rather a distasteful lesson that conservationists and naturalists have learned over a thirty- or forty-year period, in which my research is just a tiny part of the whole, increasingly showing how very disturbed the wildlife habitats of Europe are," he says. "I didn't really think it was anything more artificial than what had already happened."

He admits, however, that the large blue butterfly had a special significance for him, and that not all extinctions would have affected him so deeply. "The thing about the large blue is that one had really sweated over it for years. When it actually went, it was quite a drain."

The large blue butterfly, officially declared extinct in 1979, was reintroduced in 1983, after it was clear that no uncharted sites had been missed, that there were no survivors. Jeremy says they hadn't expected to uncover survivors, because so many surveys had been made when one population was still present. In addition, the large blue hadn't been discovered in any new regions in Britain this century.

The first step in preparing for the reintroduction of the large blue was to ensure that grazing on its former sites was adequate to encourage *M. sabuleti* to build nests under the wild thyme. This step was accomplished with support from the Somerset Trust, the National Trust, and English Nature. In areas where, just fifteen years earlier, it had been difficult to find *M. sabuleti,* acres of thyme and grassland were cultivated that supported at least one *M. sabuleti* nest per square meter.

Then ninety-three caterpillars from thyme plants in the Alvaret desert, on the Swedish island of Öland in the Baltic, were imported into Britain. Jeremy waited outside London's Heathrow International Airport as a colleague cleared them through customs. Three licenses were needed: one to import the caterpillars, one to release them, and one to work with an endangered species. At the last minute, there was a hitch: customs officials could find no regulation covering the importation of caterpillars. Lacking a regulation and a precedent, but under pressure from weary scientists inside and outside the airport facility, the officials finally admitted the caterpillars as British subjects, classifying them officially under a regulation for parrots.

These "parrots" were then released near *M. sabuleti* nests in Devon and emerged eleven months later, in late June and early July, as large blue British butterflies. Eleven generations later, in 1994, 150 adults survived on the site.

A second small introduction was carried out in 1991 on a hillside not far from the first reintroduction site in Devonshire; the butterflies' numbers have doubled there each year, reaching a thousand eggs in 1994. The greatest success, however, has been on a Somerset Trust reserve where 283 caterpillars were released in 1992 and more than a thousand adult large blues flew into the open in the summer of 1995. Scientists have taken adults and caterpillars from this reserve to start two more populations in Somerset and the Cotswolds.

Another colony has reached fifteen hundred in four generations, others are slower. Jeremy believes the more sites, the better, because he knows some will be lost. The goal is twelve to fifteen populations during the next ten years, on sites that can support more than a thousand individuals each. The Butterfly Conservation Society in Britain has undertaken to raise funds to cover the full cost of the reintroduction and maintenance of the large blue, about three-quarters of a million pounds.

Jeremy is confident that the large blues will reappear and multiply on at least four of the six sites where they have been reintroduced, in spite of droughts and other problems. But not much is being left to chance. "There is no question that the areas must be managed; they are all artificial."

More than the large blue and its red ant host are prospering under thyme in Britain. The re-establishment of the large blues' habitat has triggered some chain reactions: the pale dog-violet, along with more common violets, has increased more than a hundred fold on two large blue reintroduction sites. The increase has enabled other butterflies, the small pearl-bordered fritillary, *Boloria selene,* the pearl-bordered fritillary, *Boloria euphrosyne,* and the high brown fritillary, *Argynnis adippe,* all of which feed on the violets, to multiply. Bristle-bent grass, *Agrosis curtisli,* has spread, and the grayling butterfly, which eats the young bristle-bent, has increased at least tenfold. Other animals have also benefited. A rare bee-fly, *Bombylius canescens,* and the green tiger beetle, *Cicindela campestris,* now abound on two sites. And the woodlark, a bird that has become rare in northern Europe, breeds regularly on one of the sites.

The large blue was not the first butterfly reintroduced into Britain. The large copper, *Lycaena dispar,* another glamour species for Victorian collectors, disappeared from Britain in 1851. A beautiful, deep orange butterfly with a wingspan of forty to forty-two millimeters, it had lived in the fens of Lincolnshire and Cambridgeshire, breeding along dikes where reeds were cut, laying its eggs on Great Water Dock. These British colonies were lost forever when the fens were drained. A Dutch subspecies, *L. dispar batavus,* was introduced in 1927 and again in 1969. Permission to visit the nature reserve where the large copper now lives must be obtained from the Nature Conservancy Council.

Jeremy considers the reintroduction of the large copper to be only a partial success. He reflects that it really is a personal decision how much help nature should be given to assure the survival of a species.

He disapproves of the level of interference required to preserve the large copper, whose caterpillars survive only because a warden finds them and put cages over them. "It's one thing to take it into captivity for one last desperate effort, but I think if you can't get these things flying around freely in the wild, generation after generation after generation, even if you are doing very special things with their habitat, then you forget it. It's just a zoo," he says.

Today the World Conservation Union (IUCN) lists the large blue as "vulnerable" throughout the world, one step below "endangered." Jeremy agrees that it is probably the right classification, considering the steps being taken.

"Without active conservation immediately—and it's getting it in the northern part of Europe—it would go altogether, very soon I think," he says. "And that probably applies across Asia as well—that's in traditional land use changes. In central-southern Europe, just at the moment, its conditions are being produced in some regions quite well." He adds, though, that he'd be "extremely surprised if that carries on for another ten or twenty years, as the economic situations change." And if any less conservation were done, he says he wouldn't expect the large blue to survive as a species beyond his lifetime. Only with the programs and resources now available *can* it be kept going, under management, in Britain.

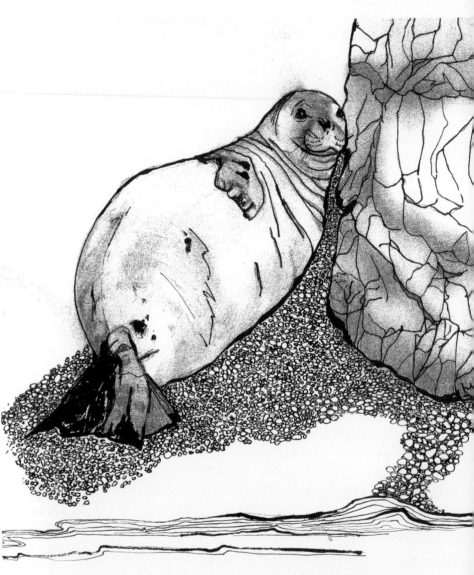

Monachus monachus. Drawing of a Mediterranean monk seal by Lorene Simms.

6

MONACHUS MONACHUS, IN RETREAT

Rights? Nobody has a right! Nature doesn't have
rights and wrongs. . . . The seals do have existence, as
we do. And who are we who govern and rule the
world?—ALIKI PANOU

 HOMER TELLS HOW the god Proteus, as the Old
Man of the Sea, emerges from the water—

and when he comes out he will sleep, under
   hollow caverns,
and around him seals, those darlings of the sea's
   lovely lady,
sleep in a huddle, after they have emerged from the gray sea,
giving off the sour smell that comes from the deep salt water.
The Odyssey, lines 404–406, translated by Richmond Lattimore

Today "those darlings of the sea's lovely lady," the Mediterranean
monk seals, hide in remote caves on islands, including Odysseus'
Ithaca and neighboring Cephalonia in the Ionian Sea. Not even Pro-
teus in all his forms could now find enough of these animals in one
place to huddle among. With perhaps only four hundred remaining,
they are among the most endangered mammals in the world, and the
rarest surviving seal.

Now, more than ever, they live up to the meaning of their Latin
name, *Monachus monachus:* monk or loner, living in retreat with no
more than one or two others. *Monachus monachus,* though still better
off than the Caribbean representative of the same genus, *Monachus
tropicalis,* which has not been sighted since the 1950s, is more threatened than the Hawaiian species, *Monachus schauinslandi,* which
numbers about fifteen hundred.

Seals of the genus *Monachus* have been in existence for more than fifteen million years. Unlike the earth's thirty other species of seals, they live in warm seas. Most Mediterranean monk seals live in Greek waters; the second largest group, perhaps a hundred seals, live off northwestern Africa near the coast of Mauritania. That population suffered a die-off in May of 1997, and it is difficult to say how many have survived. The species has virtually disappeared from the coasts of Spain, France, Italy, Portugal, Egypt, Israel, and Lebanon.

Averaging about seven hundred pounds, the adult often reaches nine feet in length, and it can live for more than forty years. Its short fur, which changes every year, varies in color from black, dark gray, or brown to light gray or beige. Jean Hermann named the seal *Phoca monachus* in 1779 when he described a skeleton brought to his museum in Strasbourg; the genus was later changed to *Monachus*. The name may originally also have referred to the animal's dark color and thick neck, which bring to mind a monk's cowl.

Monachus monachus has retreated to hidden caves often so remote that skilled divers speak with admiration of the animals' ability to find and live in them, especially those accessible only from underwater entrances. Here the seals avoid the fishermen's anger and nets as well as the developers' dynamite and wrath, a hostility born of the fear that protection of the animals will hinder construction of new beach resorts. A tangled web of government bureaucracy has been cast over the seals' fate, and numerous environmental and conservation groups compete for money to save an animal that simply wants to be left alone. The seals are victims in a modern Greek tragedy: even people with the best intentions and deepest dedication to preserving the seals can do little but count their dwindling numbers.

Aliki Panou, who has studied the Mediterranean monk seal long and intensively, is not surprised that it has withdrawn from much of its old territory and moved from open beaches into caves. "So they retreat to remoter caves and they retreat and retreat and retreat, to where?" she asks, somewhat aggressively, torn by her desire to protect the monk seal, sympathy for the fishermen who sometimes compete with it for their livelihood, and anxiety about developers, tourists, and others who infringe on its territory and even exploit it.

She does not believe that the fishermen are the only threat: the fear of humans that makes the seals retreat, she explains, is exacerbated by the explosion of tourism in the Greek islands over the past twenty years, the accompanying proliferation of boats and swimmers, and the destruction of beaches and waterfront property to make way for resort hotels.

Aliki is a forceful and fiercely independent spokesperson for the seals and the efforts to protect them, which, she is the first to say, have been largely uncoordinated and ineffective. She states emphatically that not much has been done to help the monk seals during the past decade: precious time has been lost.

Speaking on a cold day in February at the home in which she grew up, a house that was once in the country but has now been absorbed into a sprawling suburb of Athens, she describes the long and sometimes rocky path leading to the monk seals and her work to protect them. Along the way she has learned more about her fellow Greeks, bureaucrats, and the politics of international conservation than she has about the seals themselves.

Eager to be independent, Aliki left Athens, after graduating from its German school in 1972, to study biology under Jürgen Jacobs at the University of Munich, then stayed on when the possibility of a doctoral position was offered. While in Munich, she signed on to every marine biology excursion available: to the Red Sea, Chile, Kenya, and Thailand. She also worked six times as a field assistant on marine courses. When funding for the anticipated doctoral post fell through, a professor suggested that she talk with Thomas Schultze-Westrum, then also in Munich, who had developed an interest in the monk seals in Greece and who had earned his doctorate with Jürgen Jacobs' father.

Growing up in Greece, she had frequently swum, sailed, and dived—but she had never seen a seal. Only in Munich did she learn that seals lived in Greek waters. Schultze-Westrum, she explains, "is a man of fascination," and she was captivated by what he told her about the monk seals. In 1981, she traveled with him to the Sporades Islands in the northern Aegean Sea. As a student studying snakes and lizards, he had heard from Greek fishermen about the seals in the Sporades and had seen the need to protect them, perhaps by establishing a marine park for them there. But even on that trip Aliki did not spot a seal. Only a year or two later did she finally see her first monk seal, while working for English biologists who were making a survey of the islands. It was on an isolated island named Piperi, twenty miles from the closest inhabited island.

"Actually, Mother saw it first," she laughs. Her parents, unwilling to let their only daughter venture into an uninhabited part of the Aegean by herself, had decided to accompany her. They hitched a ride with a fisherman who dropped them off on the island with a tent and a promise to pick them up a week later. Swimming along a sheer coastline, they peered into every cave and cavern they could find. Her mother, a petite gray-haired woman, jokes that Aliki sent her ahead into the most difficult caves. She clearly still relishes having been able to see a monk seal pup on the beach.

On that trip and the three that followed during 1983–84, Aliki saw at least six seals. The sea mammal researchers from Cambridge then invited Aliki to join their project, supported by a grant from the European Community (E.C.) to survey the coastlines and locate suitable habitat for seals. In three areas of Greece that they had surveyed intensively, including one where Aliki accompanied them, they found nothing. Aliki shrugs and says she had not been surprised or discouraged: "I was already convinced that you can't find all the caves in ten days, in such areas."

The same biologists asked her to take part in a one-year research project, which started, after delays, in 1985. Aliki waited out the delays, because by then she was sure it was what she wanted to do. "I liked the idea, it was something with the sea, it was something in Greece and, with all those stories I'd heard from Schultze-Westrum, I liked the concept very much. I thought this was something of worth," she says.

Just what had Schultze-Westrum told her that had excited her? "The exciting thing was that he worked with the fishermen," she explains. "It was something to achieve with the fishermen, together, not excluding man from an area just to save one seal or two, or a species. It was a holistic approach." She adds, pragmatically, that it would have been impossible to launch a program without the fishermen, anyway. Schultze-Westrum had learned fluent Greek and, with the assistance of the Hellenic Society for the Protection of

Nature, the oldest nongovernmental organization (NGO) in Greece, helped the fishermen in the Sporades take steps toward forming a cooperative, with the idea of setting up a fund to help compensate fishermen who suffered from damage inflicted by the seals. "He helped them, he worked with them, appreciated them, and they agreed, and this was the big step forward at that time," she says. Thus, in 1976 fishermen in the Sporades declared officially that they were willing to protect the seals in the area, provided that nobody else was going to fish their waters and that in the future they would be compensated for losses caused by the seals. Their hope for compensation was never fulfilled.

Schultze-Westrum was also among those who proposed the area as the site of a marine park, an idea that became a reality in 1992 with the establishment of the National Marine Park of the Alónnisos, Northern Sporades.

Aliki points out that she was "just an employee" in the project with the Cambridge scientists, although she admits that having a Greek in charge of the area may have had advantages. The scientists were testing a new technique, using a camera in the caves to determine the number of seals, their size, and their sex. She was to be in charge of the Sporades portion of the survey. But when the Cambridge crew arrived to set up cameras in the Sporades in May 1985, the first major problem arose: "The Greek ministry threw us out."

It appeared that the Ministry of Physical Planning, Environment, and Public Works, which was responsible for the seals that were now drawing increasing attention from biologists and conservationists, wanted full control of the Northern Sporades.

Aliki, no shrinking violet, went to Athens to demand an accounting from the head of the ministry; she fought for eight hours to have him change his mind and allow them to work in the Sporades. In the end, he acknowledged that there was no legal basis for keeping the researchers out of the Sporades, but he would not give in. Seeing no solution in the Sporades, the Cambridge group decided to move to another area, the island of Cephalonia in the Ionian Sea. It was the first time Aliki had been to Cephalonia—and, referring to her adversary in the environment ministry, she says with a smile, "This was the only good thing this person has done, even if he doesn't know about it."

There, on the largest of the Ionian islands, the project funded by the European Community was launched, but work was hindered first by technical problems with the cameras and then by simple inability to locate any seals. Aliki resorted to counting people on the beaches as an indicator of human activity in the area. Yet she believed the local fishermen who insisted that there were seals in the area, and finally she saw one, from a distance, in northern Cephalonia. Soon afterward she also saw a newborn pup with its mother near the small fishing village of Assos. The biologists moved to Assos for the winter with high hopes and continued the study. They took an inflatable boat as close to shore as possible and then swam, dived, or rowed into every hole that looked like a cave.

Aliki spent many hours talking with local fishermen and year-round islanders. She knows that the fishermen have thought at length about the fate of the monk seal. "I have heard them say many times, 'Yes, we went into their kingdom; they didn't go into ours.'" She learned, among other things, that thirty years earlier people had sighted monk seals in the open sea, and up to fifteen seals in a cave.

By 1985, Aliki realized that strange stories about the Cambridge group were circulating. She had heard reports that the biologists were there to feed seals from their boat, or that at night they were releasing pups brought to the Ionian from a breeding station (for which plans had been reported in the media).

"And I realized that this was the worse that could happen," she says, for she knew that the public found it hard to distinguish between a proposed plan and a reality, and rumors were even harder to deal with. No breeding station has ever existed.

It is of great importance, Aliki says, that there be mutual respect and understanding between local people and researchers. In Greece, a man is expected to be at ease with the sea, and with boats. "Somebody coming out to do this kind of job, a man who couldn't repair an outboard, he would be ridiculous in the eyes of the fishermen." This makes it hard on scientists who are more comfortable with books and computers than outboard motors and skiffs.

It is different for her, Aliki notes, because she is a woman: not knowing how to repair an outboard is all right. As it is, many of the things she does routinely—handling the inflatable, diving into caves, making detailed studies—already far exceed the fishermen's expectations for a woman.

The relationship she has built with the people on Cephalonia is based on mutual collaboration, trust, and friendship. That she is Greek and enjoys the work also helps. When she goes to Athens, she gladly buys spare parts for the fishermen, and on the island they cheerfully lend her the tools she needs. She gives them credit for knowing much more about the sea than she does; she recognizes that "what we learn in university is theory."

Aliki and Jürgen, who constantly supported her, followed up the Cambridge project with another in 1986–88 on Cephalonia and Ithaca, two islands whose combined population is about thirty thousand people. With the help of her brother, Dimitris Panos, who joined the project in 1986, Aliki regularly surveyed caves, recording signs of habitation, including the "tracks" the seals make as they drag themselves over the sand and rocks: sometimes the depression where their flippers rested was obvious, other times the seal's mustache marks in the sand could even be seen. She also recorded sightings of seals and instances of damage to nets.

That fishermen are hostile to seals is well known. Deliberate killing by fishermen ranks along with habitat loss as a major threat to the monk seal population. Still, Aliki was sympathetic as she listened to stories of how the seals ate fish out of the nets and, worse, tore up the fisherman's means of livelihood. "Mainly it is the gear," she explains. "If you lose a few kilos of fish, it's not that much, but if you spend three days mending nets or you have to buy whole nets, this is the main loss." By Aliki's count at that time, 19 percent of the fishing trips from Assos encountered damage from seals. Few seals remain in Greek waters, but also fewer fish: fish taken from nets are an especially easy catch for the seals.

At one time, up to a quarter of all seal deaths could be attributed to drowning in nets, but during the past decade no accidental deaths of seals in fishing gear have been recorded in the central Ionian Sea.

Talking with fishermen on Cephalonia, Aliki realized that they would welcome compensation for their losses similar to that which Schultze-Westrum had advocated in the Sporades a decade earlier. She encouraged the men to write a letter to Hemmo Muntingh, a member of the European Parliament in Strasbourg who was particularly interested in the environment. They wrote in December 1986, describing the problem with the seals and admitting that they had no solution but appealed to the parliament to find one. One letter, signed by twelve fishermen of the fifty families in the village of Stavros Ithaki, emphasized that they lived only from fishing, which they had practiced their entire lives. "The seals in that area survive because we do not want to kill them," they wrote. "We are asking for a solution to the problem as fast as possible." Copies of the letters were sent to the Greek government.

The results? "From the Greek government, nothing except some money to repair the dock in Ithaca four years later," says Aliki. Hemmo Muntingh responded by visiting the fishermen in 1987. He explained how the European Parliament works, that Greece first must propose a solution before the European Community could act. Aliki shows her impatience with the bureaucracy in a simple observation: "Fishermen don't care about these things; they want to see something done. They cannot make proposals."

Fishing is declining as a profession in Greece, and the men who fish these waters are few. The population of the fishing village of Assos on Cephalonia, for example, is about sixty. The story is similar throughout the Greek islands: those who do not fish must increasingly make a living from tourism, which Aliki considers a greater threat to the seals than fishing, for it results in more disturbance and loss of habitat for breeding and resting.

Aliki began to take children from the villages to see the seals' caves. She and Dimitris weighed the risks of showing the children the caves and decided to go ahead, in the belief that involving the children and educating them about the seals were the best things to do. The children might have discovered the caves anyway. "If we do things in secret, this is worse." Going in from the water in an inflatable boat with no engine, they would pull themselves along on the roof of a cave until they got to the protected beach inside. The children were eager to explore the caves and were amazed at what they found. During the last two months of 1995 alone, Aliki and others made twenty-four presentations to twelve hundred schoolchildren on Cephalonia and Ithaka about the plight of the monk seal.

Aliki often talked to the children about a young seal, which the biologists had named Penelope, after the wife of Odysseus. According to Homer, Odysseus' faithful wife lived on Ithaca; Penelope the seal lived in the Ithaca Channel and was faithful to her cave. Aliki took the schoolchildren to visit the seal. When the Cambridge crew looked at photos of Penelope the seal, they determined that Penelope was, in fact, a male. Nonetheless, the name stuck. Once, when it was not possible for Aliki or Dimitris to make a survey, they sent a friend in their boat. When the stranger entered the cave, the startled Penelope began to flee, then recognized the familiar boat and stayed. Other seals have come to recognize the sound of engines, to distinguish between tourist boats—whose occupants often lavish attention on them—and fishing boats, whose occupants may shout or shoot at them.

Penelope has disappeared; no one knows why. Few other seals have been seen in the Ithaca Channel since Penelope left.

An information network was set up so that local residents could call in to report sightings of seals, a sick pup, or other matters of particular interest to the biologists. Aliki had decided early on that accompanying fishermen on their boats would not be worthwhile because sighting a seal at a net was a relatively rare event. Living and working out of an ancient Venetian lighthouse on Cephalonia, her use of which was sanctioned by the Hellenic Navy as its contribution to conservation, she talked every day with the thirteen fishermen on the island of Ithaca and the five or six on Cephalonia.

She has foregone gathering data wherever there is a danger of disturbing the seals; even so, between 1986 and 1988 she recorded nearly four hundred reliable sightings, representing a total of between eighteen and twenty-five seals. The population in the area appeared to be stable. Every year, two or three new pups would be seen, and every year a minimum of eight to ten seals were recorded. But Aliki is careful to acknowledge that those seals may or may not be the same ones sighted the year before. No attempt to photograph or tag them has succeeded. Because they move from cave to cave and from island to island, they cannot be identified by location. Size and color seem the best indicators, but it is very difficult to see detail from a distance. For those reasons, no one—including Aliki—knows how many are left. The solution, she believes, may be telemetry—gluing small very high frequency radios and antennas to the animals' skin. This has not yet been done with monk seals in the wild.

The report on the 1986–88 project, published in 1993 in *Biological Conservation*, acknowledged the need for more research on the ecology of *M. monachus:* "Data on rates of natality and mortality, age structure, home range and dispersal are missing." The report also called for an end to deliberate killing, compensation to fishermen for losses caused by seals, and the establishment of protection zones.

Aliki and Jürgen applied for money to continue. In September 1988, World Wildlife Fund International accepted their proposal to maintain public awareness of the monk seals and to collect data from a larger area. The Greek government gave the proposal a green light in May 1989.

Two years later, Aliki was asked to prepare the budgets and plans for a three-year WWF project. Philippe Roch, then with WWF International in Switzerland, had urged that the project be expanded into an integrated effort for the central Ionian Sea, to include, with regard to monk seals, physical planning, promotion of protected areas, and information about alternative sources of income for residents. Aliki, who thought that the idea of an integrated project was crucial, was enthusiastic. She believed that before any restrictions were put in place, however, the negative effects should be taken into account and remedied.

About this time, WWF Greece was established, and supervision of the project was transferred from the international organization to the new director in Athens, Georgia Valaoras. In 1991 Roch and seven other WWF administrators, including Valaoras, visited Aliki on Cephalonia and approved the concept of the integrated project. In February 1992, a "country team" of eight WWF members, this time including Magnus Sylven, director of the Europe and Middle East Regional Program, paid a visit to Aliki and also voiced support for the concept.

Then, against the recommendations of Aliki and members of the country team, WWF Greece decided to expand the concept to a project valued at two million European Currency Units (ECUs). This would include three additional subprojects on the island of Zákinthos (pertaining to seals, turtles, and the purchase of a turtle beach), birds on Strofades south of Zákinthos, and turtles on the Peloponnese. This concept was approved by the European Community.

Aliki changed the objectives and the budget six times within six months, finally drawing up a plan for the central Ionian Sea that restricted the integrated project to five points whose inclusion was most urgent for integrated management of the area: continued monitoring of seals, monitoring of incidental catches of turtles in swordfish long lines, development of a management plan for the area, an education package for schoolchildren, and coordination of the various parts. This fifth point, she knew, would be the most difficult to accomplish.

Although the project was officially launched in July 1992, Aliki, who still had no contract, was notified in October by WWF Greece that the funds that had been committed to her project would be cut because WWF management fees had not been taken into consideration. Moreover, a program to promote ecotourism through self-guided walking trails, which had been under a separate WWF contract, would be added to Aliki and Jürgen's project.

Aliki and Jürgen traveled to Athens to question the need for the changes. The portion of the grant allotted to Cephalonia had already been cut to less than 15 percent of the whole, and now further reductions would take place. Aliki well remembers the meeting with the director: "She said we forgot this and that and that, you will get less and less and less." The monk seal segment, which had been the impetus for the "integrated" project, had been downgraded to a minor role. "And then we said, 'This is ridiculous.' That's when Jürgen screamed. This was the only time I saw Jürgen scream," recalls Aliki of the man who had supported her since 1976 and who had always cosigned the grants.

Their protests were to no avail.

They were not frustrated merely with the finances, the bureaucracy and reorganization of the project. Time that was precious for the monk seals had been lost; no progress was being made. It was especially discouraging because the WWF International commission that had come to Cephalonia had decided that Aliki and Jürgen's monk seal project was an ideal precursor for an integrated project. They had

praised both its organization and the islanders' acceptance of it. She recalls also that when WWF International collected money for the monk seal project in Switzerland in 1990, "the Swiss people gave . . . double what was needed." Philippe Roch had then urged Aliki to speak up if she needed anything. She had done so, providing plans and budgets to extend the monk seal project to the island of Zákinthos.

After her protests of the additional cuts were overruled in 1992, Aliki was forced to retreat, but she refused to give up. Bowing to the WWF bureaucracy in Athens and faced with limited funds—not a new situation—Aliki worked within the reduced budget for the next three years. She continued the project on Cephalonia, putting out leaflets on monk seals and giving lectures on turtles. Her assessment is that the three-year project, which ended in June 1995, made no real progress, a situation "almost worse than zero." Not only did the monk seals lose out, but the public lost faith in the project, especially after the official announcements of the five-point Ionian Sea project had already been made at the local level.

The WWF monk seal team on Zákinthos changed several times during the project and was hampered by disagreements and friction among the workers. In August 1995, Jürgen Jacobs, Aliki's friend, mentor, and constant supporter of nearly twenty years, died of cancer in Munich. In the winter of 1995–96, Georgia Valaoras left WWF Greece. In the end, no one was left to carry on the monk seal research except Aliki and a few part-time workers.

The WWF project was followed by a six-month grant directly from the European Union. Under funding regulations, the European Union pays only up to 75 percent of the project cost; the remaining 25 percent is to come from local sources; but when no local funds are available, either work stops or the workers volunteer their time. Aliki is philosophical about becoming a volunteer: "It is the first time without being paid," she says simply. "What keeps me is that I am doing something I like, in an area I love, with people I love, something I believe is correct and good. And I can more or less do what I think."

In September 1995, fishermen on Cephalonia wrote another letter, this time to the Department of Fisheries, under the Greek Ministry of Agriculture, pleading for a response to their request for a solution to the problem of damage inflicted by seals. They urged the government to implement a proposal to evaluate damage at the national level. Such a proposal had been submitted within the framework of the PESCA program developed by the European Union. No PESCA program has been implemented in Greece, Aliki says, because the Ministry of Agriculture has not been able to decide how and by whom it should be managed. The fishermen, Aliki stresses, feel that "it is not correct and right, ethically and practically, that the whole of Europe wants to preserve one animal and the cost is paid by thirty thousand fishermen." And, she adds, "This is my opinion, too."

The Greek government made no response to the letter.

Aliki's strongest criticism of the government is that it has responded either not in time or not at all. She regrets the lack of continuity, both in what is being done and in the people carrying out the work, and the lack of a strategy for the conservation of the monk seal. The projects that are launched are uncoordinated with—or even contradictory to—each other, and the priorities change every year or so. First "it is captive breeding, then it is protection *in situ,* then it is that and that and that."

Too little interest is devoted to the seals in some quarters, and too much in others. In the impressive *Book-Directory for the Mediterranean Monk Seal in Greece* published in Athens in 1994, at least ten groups present their work. With so many different agencies, ministries, NGOs, and nonprofit groups all wanting to save the monk seal, coordination is obviously difficult. Aliki refused to contribute to the directory, saying it was a waste of time and of the fifty thousand ECUs it cost to produce it. She was too busy with the monk seal work to write another report.

She is cynical about many of the groups, and about the people who head them. "My personal opinion goes so far that I have said quite often to friends, 'I think they are recycling their own jobs.' Because it is impossible for me that in ten years there's no progress. So what did they do in that time?" The trend, she says, is obvious: yuppies from Athens, who don't know the fishermen's mentality or concerns, predominate. Meanwhile, WWF Greece has acquired a reputation for being an arrogant NGO that ignores other authorities, according to Aliki.

When Aliki was asked in 1992 by the national coordinator of

monk seal projects to comment on a proposal to the European Union, she initially declined, saying that no one would pay attention. On being urged by the coordinator several more times to comment, she wrote out an opinion of the project, point by point. To make things clear, she drew up a five-page strategy draft that could be used later to make the official proposal. "And he didn't read it at all; it didn't matter," she sighs.

She describes an all-day meeting that took place on March 18, 1994, between the E.U. representative and the numerous groups working on the monk seals. When her turn finally came to speak, at 7:00 P.M., she asserted that she would not present any results from her work but would instead discuss the problems with the whole program, beginning with lack of coordination and strategy, at both the E.U. and the Greek level. The Greek coordinator exploded, she says, but the E.U. representative accepted the assessment. "That was the beginning of the end of those ridiculous things," she concludes.

The "ridiculous things" included what Aliki describes as "cooking the same soup all the time"—that is, applying for grants for one, two- or three-year projects that rely on methods already demonstrated to be worthless. Another of the "ridiculous things" was a self-nominated national coordinator who oversaw various organizations that were "coming up like mushrooms because of the money they smelled."

The Greek ministry then refused to take responsibility for choosing among the various groups and declined to support any proposals. And the European Union, for its part, would accept no proposal from Greece unless it was supported locally. Aliki views with great frustration the loss both of the time and of two years' worth of money from the European Union for the monk seals.

She is convinced that the fate of the monk seal depends on the individuals in public organizations or in other groups such as NGOs that are in a position to influence conservation movements, administer funds, and set policies. She believes that although the Greek Ministry of the Environment was at one time an obstacle to helping the monk seal, that is no longer the case, primarily because the personnel has changed over time. Her determination to persevere with her efforts to save the monk seal appears unshaken. She continues to do research and to combat both habitat destruction and deliberate killing, which she thinks are equally problematic but have to be dealt with separately.

As for conservation in general, however, she believes the root of the problem—the demand by Western countries for natural resources —is not being addressed.

In her characteristic no-nonsense way, she also condemns the arrogance of people who are quick to point the finger at others. Global population growth, she believes, is a huge problem. But of the Earth Summit in Rio de Janeiro in 1992, she asks: "What came out of it? The wealthy countries said to the Third World, 'You're not allowed to have a car, because I already have two.'"

Aliki has thought about how she would feel if one day she went looking for monk seals and couldn't find any, and if that continued day after day, week after week, until she realized that they weren't just in retreat but had disappeared forever. "If they go extinct, which is probable, it should at least serve as a lesson [for] why we didn't succeed," she concludes sadly.

In 1996, at least one seal pup was born, and probably two. At the end of 1996, Aliki was still not gainfully employed on any monk seal project. "That is okay with me," she told us. She has by no means stopped monitoring the seals or working on educational activities. In spring 1997 she registered in Germany to begin a long-delayed doctoral degree on the Mediterranean monk seal, for which she has collected more than ten years' worth of data.

She continues her conservation efforts under the auspices of a local NGO she helped set up called Archipelagos—Environment and Development. Archipelagos is a private, nonprofit conservation group headquartered in Cephalonia. Protection of the Mediterranean monk seal is regarded as part of its more far-reaching attempt to protect the environment of Cephalonia and Ithaca, primarily by developing a plan in which tourism, nature, local culture, and the economy complement each other. Aliki's overhead is low, her lifestyle simple. She still lives on Cephalonia in the old Venetian lighthouse belonging to the Hellenic Navy, for whose continued support she remains grateful. From here she can quietly monitor the monk seals, contemplate other conservation projects, and work on her doctoral thesis.

Asked if she thinks the monk seals have rights, she declares: "Rights? Nobody has a right! Nature doesn't have rights and wrongs!" Then, more calmly, she adds: "The seals do have existence, as we do. And who are we who govern and rule the world?"

As she tells the fishermen: "The sea can provide enough food for everybody—fishermen, dolphins, seals, everything. That's the point. If the seal goes extinct, it's a bad indication. We cannot ask what is the benefit of the seal, which is the most frequent question." During World War II, she explains, islanders relied on the seals, to provide soap and shoes. Today too few seals remain even for that.

This chapter on the Mediterranean monk seal was originally to have included two other interviews held in Greece. We have decided that neither of those interviews can be included, for reasons that reveal a lot about the situation.

One interview was with a man who had keen insight both into environmental issues concerning endangered species and into the highest levels of environmental politics and academics in Greece.

We were impressed by his candor and his willingness to speak in detail about how money meant for conservation was diverted to personal ends. He talked about how national conservation organizations were being turned into fronts for private interests and how E.U. policies had created a structure of incentives and disincentives that were destroying traditional Greek culture—and with it the Mediterranean landscape and its animals and plants, both on land and in the water. His outrage and integrity shone through the interview.

Six months after that interview, we received an express letter from him asking us to delete all mention of him from our book. Publication of his opinions, he wrote, would cause him "serious embarrassment" and make it difficult for him and his family to continue to live in Greece.

The other interview we had hoped to include was with a scientist who has spent many years studying the Mediterranean monk seal in Greek waters. His story bore many similarities to Aliki's: lack of cooperation, even hostility, from Greek authorities; sporadic funding

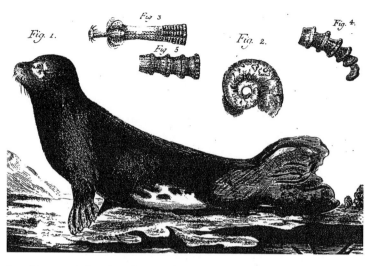

Jean Hermann's drawing of *Monachus monachus,* the Mediterranean monk seal, which he originally christened *Phoca monachus* in 1779, when he described it from a skeleton. The skeleton can still be seen in the Musée Zoologique de Strasbourg, which Hermann founded. Illustration from J. Hermann, "Beschreibung der Mönchsrobbe," *Beschäftigungen der Berlinischen Gesellschaft Naturforschender Freunde* 4:456–509.

for his work; difficulty in finding the elusive seals; strife and mistrust between conservationists and fishermen. In addition, he related the disturbing experiences he had had during 1991–93, while working on a monk seal project with a well-known conservation organization—from the friction between the local people and the conservationists to the necessity of concealing his observations of monk seals. He knew that if he could see the seals and make videos of their feeding behavior from high cliffs, other people could observe and kill them from those same cliffs. In fact, when a seal was killed by dynamite during that time, he was convinced that it had been targeted by people who wanted to build a tourist development on a beach used by the seals. To drive home their point, they sent him the autopsy report on the seal. He told us how, in the spring of 1993, a seal was shot by a fisherman from a boat in broad daylight. This provocative act was witnessed by locals, and it divided the members of his conservation group. Some wanted to prosecute the fisherman;

others opposed taking action because they feared reprisals against the monk seal project. In the end, the fisherman was never brought to trial.

The most disturbing thing he told us, however, concerned the infighting within the conservation group he worked with—and the jockeying by some of its members to replace the project leaders with their own friends from Athens. He charged that the project had been sabotaged by workers in the organization's headquarters, that dangerous conditions had been created by the organization's refusal to repair or replace boats or equipment, and that he and the project leader had been forced out of their positions and told that their data must be turned over to their replacements. When he left the project, after refusing to give up his raw data but promising a full report on his work, the hard disk on the project leader's personal computer was destroyed. Some of his own data on that computer, he added, were stolen by the workers who replaced him. He vowed that he would never again work for that conservation organization. He has since founded his own environmental firm and has continued his work on monk seals.

Some time after we wrote about his experiences, a woman at the conservation organization's headquarters in Athens got in touch with us. She had heard about our interview and wanted to tell us her side of the story. She and her "companion in life at the time" had been directly involved in the story we had been told, she said. We read faxes from Athens and weighed her charges and countercharges, many of which had to do with misuse of funds. Because we could not determine whose version we could trust, we decided to eliminate that part of the story. We did so reluctantly because we had believed the researcher we interviewed, but faced with another version of the same events, we had no way to confirm much of what he had told us.

One thing is very clear: conservationists in Greece have been doing just about everything except saving the monk seal, which appears almost as a footnote to the infighting and maneuvering of the people paid to protect it.

Newspapers recently reported from Athens that more than forty environmental groups there have urged emergency measures to save

Mediterranean fish populations from extinction. Considering what we have heard about the administrative and social problems within and among Greek conservation groups, we are pessimistic about the outlook for the fish. Perhaps their best bet is to use the "monk strategy" and go into hiding.

MONACHUS MONACHUS, IN RETREAT

Haplochromis. Drawing of a haplochromine cichlid by Lorene Simms.

7
INTRODUCING THE NILE PERCH

If I told you that there wouldn't be sparrows anymore
tomorrow, you wouldn't believe me, would you?
—FRANS WITTE

There came a sad moment when I realized that most of
these species are extinct. It gave me the feeling that
dissecting these fish is like cutting up a Rembrandt to
learn what paints he used.—KEES BAREL

For us Africans here, you will talk of Darwin's finches,
but we do not know them. But surely, when you talk
about haplochromines, we understand what you are
saying.—SYLVESTER B. WANDERA

 FROM AN AIRPLANE at nine thousand meters (thirty
thousand feet), Lake Victoria looks like a vast sea
on the surface of Africa, its waters shared by
Kenya, Uganda, and Tanzania. With a surface area
of sixty-nine thousand square kilometers (26,500
square miles) and a mean depth of forty meters (130 feet), it is the
earth's second-largest freshwater lake and the largest tropical lake.
Until the 1980s, it held an incredible quantity and diversity of fish, including a flock of many species of small, colorful haplochromine
cichlids that occurred only in that lake. Scientists use the term *flock* to
describe closely related species that have descended from a common
ancestor, then specialized on different ways of life. This genus of
small, perchlike fish was estimated to contain between five hundred
and a thousand species in Lake Victoria alone; in all of Europe there
are no more than two hundred species of freshwater fish. Scientists
had long wondered at the rapid speciation and adaptive radiation dis-

played by the haplochromines of Lake Victoria, which is unrivaled among vertebrates.

In 1954 and again in the early 1960s, in an attempt to increase the food value of Lake Victoria's fisheries, a new fish was introduced into the lake. This fish is called *Mbuta* in Uganda and Kenya, *Sangara* in Tanzania. For scientists, however, the identification still has to be confirmed. It may be *Lates niloticus, L. macrophthalmus* or *L. longispinis*. All three species occur in the lakes of its origin. It may be a hybrid. We refer to it simply as the Nile perch.

There had been some opposition to the introduction of the large predator, but it was not loud and it was not heeded. Some people thought that those who objected were short-sighted; the Nile perch would surely yield more useful protein for the people living along the shore than the little, bony haplochromines that people in Uganda know as *Nkejje* and those in Tanzania and Kenya as *Furu*. For more than twenty years not much appeared to change. Then at the beginning of the 1980s, earlier in the northeast, later in the southwest, an explosive increase occurred in Nile perch, accompanied by a decline in the stocks of the other fish, most dramatically the haplochromines. By the mid-1980s, scientists who had watched the situation were describing the complex ecosystem of Lake Victoria as "irreversibly destroyed." By 1993, they estimated that up to two hundred species of the endemic haplochromines had disappeared.

Sitting in his laboratory at the University of Leiden in The Netherlands, Frans Witte carefully reconstructs what happened; he has followed the events since before he finished his university studies in 1977. As he speaks, some brightly colored haplochromines from Lake Victoria swim in aquaria nearby. They no longer swim in Lake Victoria—only in aquaria in a few European and American cities. How many species disappeared before being collected, named, and studied is unknown.

Frans grew up in a village in Surinam in South America and has always had an interest in nature and animals. But he decided to study biology, he says, not only because of that interest in nature but "partly because I had in mind something very idealistic, I wanted to do something useful, perhaps solve the world's food problem, I don't know what."

In 1975 he and his wife, Els, were students working on haplochromine cichlids at the University of Leiden, when their teachers Kees Barel and Gerrit Anker went to collect live fish on Lake Victoria. That trip and consequently the whole project that developed out of it, had been triggered by a television program Gerrit Anker had seen that showed Holland's Prince Claus aboard a small vessel on an African lake. It was not the prince but what the boat was hauling in that had caught his eye: huge quantities of haplochromines. Excited, he contacted Kees Barel who, in turn, called the government information agency the next day and determined the prince was on Lake Victoria. The two biologists flew to Tanzania, where they soon learned about efforts to make fish from the lake more useful to the local people. A survey of the lake had showed that more than 80 percent of its fish biomass was composed of haplochromines, which were small, generally three to ten inches long, and bony. The rest comprised thirty-eight species belonging to eleven other families. Two indigenous tilapiine cichlids were larger and tastier than the haplochromines and the main target of the fishery on the lake. Local experiments with canning, drying, and smoking haplochromines had run into problems that made processing costly. Finally, it was suggested that better use could be made of haplochromines if they were ground into fish meal. Although this would be profitable, fish meal could be used only as fodder for pigs and chickens, not for direct human consumption.

The fish-meal factory, Frans says, was designed to take sixty tons of fish a day from the mouth of the Mwanza Gulf in southern Lake Victoria. Such an intensive fishery operation, in an area where the ecosystem was little understood, could result in overfishing. Kees Barel warned at the time that the proposed fishery might endanger some species, that they might even become extinct. He obtained a grant from the Netherlands Foundation for the Advancement of Tropical Research (WOTRO) for a project to collect data for the proposed fishery, the results of which were promised to the government of Tanzania.

Frans describes his interest in another goal of the project as "very selfish": to study the ecomorphology of the fish. They could study the structure of the fish in the lab, but they could not interpret its ecological significance properly until they knew something about the fish in the lake. "We had to know what the fish were doing in nature and not just deduce it from the stomach contents and from their pharyngeal jaws," he explains.

A third goal was to make a taxonomic inventory of the fish, for they kept in mind Kees's concern that fish could become extinct as a result of intensive fishing. "And we never knew at that time how right he would be, and how important the inventory would be," says Frans.

Thus, the Haplochromis Ecology Survey Team (HEST), set up by Kees Barel, started what was to be a three-year field program in Tanzania in July 1977. The program, which was subsequently extended until March 1992, employed on a continuous basis biologists from the University of Leiden, who carried out forty-five person-years of fieldwork over the course of the project.

Together with Martien van Oijen, who later became curator of fishes at the National Museum of Natural History in Leiden, Frans and Els formed the first team. Like Kees and Gerrit before them, they spent two months at the Natural History Museum in London studying the fish collected mostly by its curator, Humphry Greenwood, an expert on fish taxonomy. That collection, of about a hundred species, remains the most comprehensive assembly of haplochromines from Lake Victoria. Frans found it an impressive challenge to recognize and remember them.

Although species are often defined as consisting of organisms or populations that are capable of interbreeding, in most cases biologists can tell that two or more species are present in a given place if they remain separate over an extended time and do not blend into a single population through mating. Examples of different but similar species are wolves and coyotes, humans and chimpanzees, and bald and golden eagles. These are just six species. Lake Victoria contained a species flock of five hundred to a thousand species, similar to one another in that they were all medium to small fish, but amazingly diverse in their patterns of feeding and reproduction and their coloring. It took time and patience to sort them out. Frans and Els realized later how little they had seen in London.

Arriving in Tanzania on the shores of Lake Victoria, they were lucky to find Teunis Ras, a Dutch master fisherman, who taught them some basics: how to handle a boat and a trawl, how to get nets and how to use them. "We were beginners who knew something about the taxonomy of haplochromines but nothing about fishing gear,"

Frans explains, adding that not long after they started fishing, they had their first big surprise. "We tried to identify the species, but every time we began, we said, 'We are not the right persons to do this work'—because we could not identify them." From the structure of the jaws it was possible to identify the trophic groups (categories defined by how and what the fish eat) but that did not help when it came to identifying species. Again and again they ran in vain through their identification key. In the end, they had to give the fish they collected nicknames or numbers while they awaited proper identification or the conclusion that they were dealing with a new species.

Puzzled about why they didn't find any of the Greenwood species, they moved to the areas near the shore of the lake and fished with gill nets and beach seines. This was the main habitat Greenwood had fished, those were the kinds of nets he had used—and here they found the species they had studied in the Natural History Museum, plus many more. On Sundays, their day off, as they climbed over rocks along the lakeshore, in an area where Greenwood had found only one species, *Haplochromis nigricans,* they found an astounding number of colored fish they didn't recognize—mainly algae scrapers, but other trophic groups as well.

Lake Victoria is about twice the size of The Netherlands. The HEST group wisely decided to confine its sampling to the Mwanza Gulf, sixty kilometers (thirty-seven miles) long and five kilometers (three miles) wide, and the larger Speke Gulf to the east. "As we discovered that each trawl shot contained new species, we had to further limit our study area, since a taxonomic inventory was not the major goal of the research," explains Frans. Cutting back on their ambitious plans, they concentrated on the northern part of the Mwanza Gulf, then reduced the area even further, to a five-kilometer-long transect from one side of the gulf to the other. They placed twelve sampling stations on the transect. They had settled for sampling about one-tenth of 1 percent of Lake Victoria.

"At first it was dazzling," Frans recalls. "One hour or three-quarters of an hour brought in a ton of fish." He points to a photo taken in 1977, showing the deck covered with thousands of small fish. "In such a catch, if they did not take too great a depth range, there were about forty species. If they took a greater range, there were more." He pulls out a "dirty book," spattered with lakewater and fish scales and faded by the sun. In this field book, prestamped with outlines of fish, he and Els recorded the details, including the sometimes strikingly beautiful colors, of each species. He has a dozen other such books. Martien van Oijen also has several.

"In this way, the first year passed, until Martien, Els, and I got frustrated and said we hoped we wouldn't find anything new," Frans recalls. "Discovering new species might have been exciting to other biologists, but we had to invent a name, we had to describe the colors and other characteristics of all those new taxa." He explains that producing descriptions that can be published in scientific papers is so time-consuming that the three could do it only for comparatively few species. Only about twenty species are described in the HEST literature.

Along with this discovery of a plethora of species came two baffling questions: How could there be so many species, and how could they coexist?

"How can it be that you come in an area where so many people are living and you find a hundred and fifty new species within one year?" Frans wondered. "We couldn't believe it, sometimes we didn't trust ourselves. Are we wrong? What is going on? One hundred and fifty new species? It does not sound very nice, but at times we got fed up with it, because we were there to do ecological work."

After two and a half years, by 1980, they started to develop a picture of the complex fish communities sampled along the transect. One thing was clear: the species were very locally restricted. In trawler hauls from stations over sand the catch was almost totally different than in hauls from stations over mud. In hauls from different depths, the catches were also different, as were the hauls made at night and those during the day at the same depth.

The haplochromines were fascinating to scientists interested in how the head muscles and bones of different fish suited various feeding patterns. Virtually all food sources—algae, insects, mollusks, plankton, prawns, fish and lake-bottom debris—were exploited by the fish. Some species had such extraordinary feeding habits as stealing eggs and hatchlings from mouth-brooding females or scraping scales or parasites from other fish. Their reproductive habits also differed greatly. Some species bred throughout the year, others only seasonally; some in shallow water, some in deep water; some on sand, some in the rocks.

Keenly aware of this amazing diversity and number of fish, the scientists of HEST were quick to reflect on the impact of the new fishery, which would grind all this diversity—scientifically recorded or not—into fish meal. They saw the trawlers from the fish factory coming into areas where fish were spawning at the end of the rainy season, into huge aggregations of fish. They warned that it could go wrong, for they knew the trawlers were catching the brooding females, which keep young in their mouths for about three weeks. The trawlers also damaged the nests where the males display. The unrestricted trawl fishery posed a great threat to the haplochromines.

In the first paper Frans wrote on this work, he included warnings about trawl fishery, saying that meshes used by the fishermen were so small that they were even catching many juveniles of some of the smallest species. "Of course, you have a big problem," he says now, "because this is a multispecies fishery, and the haplochromines at that time ranged from six centimeters for the smallest adults up to twenty-five centimeters for the largest."

The big fish disappeared very fast. Frans later estimated that the trawl fishery that began during the 1970s in Tanzanian waters reduced haplochromine catch rates by more than half within six years.

Not all the fish caught at that time in Lake Victoria were going into fish meal: some were being eaten, both near the lake and in the bush. "The lake people could fish themselves, and they would choose the better fish, the tilapiines and catfish," Frans recalls. "The trawler from the Fisheries Training Institute fished for a couple of hours in the Mwanza Gulf and they would catch three tons of haplochromines, store them in fish boxes, put some ice over them, and take them into the villages. They sold the fish for only one shilling per kilogram, which was the price of an egg. This appeared a proper way to help the poor people in the villages, far away from the lake, where otherwise no fish would come. And it worked well."

Here, then, was a way to help solve the food problems of poor people, a problem Frans had long pondered. Unfortunately, it was at the expense of an incredible natural diversity of fish species that had come to fascinate him.

Plans to start a similar project near Bukoba on the western shore of the lake had been spoiled by the outbreak of war between Tanzania and Uganda in 1978. When the ice machine in Mwanza broke down and then the van for transporting the fish stopped running, local people at Mwanza began to dry haplochromines and sell them for human consumption. On a relatively small scale, they also ground them in a corn mill to make chicken feed.

In 1982, with extra funding from the Dutch Ministry of Foreign Affairs, the HEST scientists, working with counterparts in the newly formed Tanzania Fisheries Research Institute (TAFIRI), began to examine how so many fish species could coexist, as well as more applied questions, such as how to avoid overfishing. It appeared that the immediate threat to the dazzling diversity and quantity of haplochromines was over.

"And just when we arrived at that point, in 1982, the Nile perch came up," Frans announces with a sigh.

He knows the history well. In 1927–28, when Michael Graham had made the first fishing survey of Lake Victoria, he recorded fifty-eight species of *Haplochromis* and commented that "the number of individuals is almost incredible." While Graham regretted that the enormous haplochromine population was not really useful, he warned against introduction of a large predator that could convert these little, bony fish—which the colonial fisheries officers called trash fish—into large fish that could be caught for food. The leading candidate was the Nile perch, which was tasty, grew to over six feet (about two meters) and two hundred twenty pounds (a hundred kilograms), and already lived in Lake Albert. At that time Graham wrote: "The introduction of a large predatory species from another area would be attended with the utmost danger, unless preceded by extensive research into the probable effects of the operation." In a footnote, he added that his warning had just been strengthened by a recent research report from Lake Albert which described how rare the tilapia there had become.

The warning evidently did not register or was forgotten. In August 1954, Nile perch from Lake Albert in Uganda were introduced into Lake Victoria by J. Ofulla Amaras, a fisheries officer in Kenya, after clearance by the fisheries department in Uganda. In 1959, the first Nile perch were caught near Jinja, Uganda, and the debate began on whether more should be introduced. One early fierce opponent, Geoffrey Fryer at the East African Fisheries Research Organization in Jinja, issued a strong warning about the potentially disastrous consequences of introducing the Nile perch. After a brief, fierce debate in scientific journals and the press, little more was heard about the Nile perch. In 1962 and 1963 it was introduced again, from both Uganda and Kenya. In 1972, for the first time, it was caught in the southern part of the lake, near Mwanza in Tanzania.

Frans and his wife lived in Tanzania from 1977 to 1982; two of their children were born there. When the project was extended in 1982, new researchers from Leiden went to Mwanza, and Frans returned with his family to Leiden to take over Kees Barel's role as project coordinator.

Before his return, Frans noted some increase in Nile perch in the southern part of the lake. But since the collapse of the East African Community in 1977, collection and transport of fish in other parts of the lake had been difficult. In 1984, after borders had been reopened between Tanzania, Kenya, and Uganda, Frans traveled to the Kisumu Institute in Kenya, in the northeastern part of Lake Victoria. Kisumu, he recalls, was one of the places where the Nile perch was first introduced. There he asked the people if he could collect some haplochromines.

"And they started to laugh! 'What are you thinking? There are no haplochromines here!' Well, at that time I had been working for more than seven years in Mwanza and as soon as you put a net into the lake you caught ample haplochromines. To me, there seemed something wrong with their fishing technique. 'Okay,' they said, 'join us on the boat.'"

After two trawls of an hour each, they managed to catch just two haplochromines. Frans draws photos out of his desk: two small fish lay on the deck among many larger ones. "Those are Nile perch," he says, pointing to the larger fish. "So virtually the only thing we caught by trawling was Nile perch."

He shakes his head. "There was nothing, only two individuals where there used to be tons. And the trawler there even fished with a much finer mesh than we used in Mwanza. While we were using nineteen-millimeter [three-fourth-inch] mesh, they were fishing with an extra layer of five millimeters [one-fifth inch] or so, a mosquito seine, and still we didn't get anything. That was the first shock, the first time we realized that something was going on."

He then recalled that in 1981, as an observer at a meeting of a subcommittee of the Committee for Inland Fisheries in Africa (CIFA) sponsored by the United Nations Food and Agricultural Organization (FAO), he had heard a report from the Kenyan delegation that the number of haplochromines was decreasing. At the time he did not appreciate the gravity of the situation; not until his Kenyan experience in 1984 did he really begin to worry.

"And then we got alarmed and we realized that the same thing that had happened in Kenya might hit the Tanzanian area as well. But it was very difficult to believe that something always around you will not be there anymore. It is so—so shaking, you can't imagine it."

Deeply concerned, in 1985 he helped organize a "floating laboratory" on a ferry with support from the foreign affairs ministry and made a seven-week inventory in the southern part of the lake. He appealed to the foreign affairs ministry not by referring to an ongoing loss of biodiversity—the term was not yet popular—but by saying he was investigating the disappearance of food for the Nile perch.

In their seven-week inventory, the HEST and TAFIRI scientists found that the haplochromines were almost gone from the Speke Gulf. They moved to the west, into the Emin Pasha Gulf, and were still able to catch haplochromines in large quantities, while Nile perch catches were relatively low there. A year later when a colleague sampled the area, however, the number of Nile perch had doubled, and haplochromines had decreased by a factor of five.

A picture emerged of the Nile perch population as first exploding in the northeast and then sweeping clockwise down to the southwest. They also saw another trend. "When we fished in deep water, the Nile perch were more abundant than in shallow water, which was unexpected because it was always said that the Nile perch was a shallow-water species."

Frans calculates that by 1985, Nile perch dominated the fish fauna in most of the lake. In 1987, they caught only fifteen haplochromine individuals on the transect across the Mwanza Gulf. In 1988, they found nothing but Nile perch and *Rastrineobola argentea*, a small pelagic cyprinid fish, locally called *Dagaa*.

"It seemed that Nile perch were migrating in a wavelike movement through the lake. With the meshes used for our haplochromines we should have caught juvenile Nile perch if the initial increase was the result of spawning in the Mwanza Gulf, but the first invasion consisted of subadults and adults," Frans says. The subadults, he adds, were at least twenty-five centimeters (ten inches) long. In only one to two years after this invasion, juvenile Nile perch smaller than ten centimeters (four inches) appeared in the catches.

He remembers how "strange" things became then. In Kenya, where the Nile perch population first exploded, it did so in shallow water: densities in deeper water were lower. The haplochromines had complementary densities: lower in shallow water and higher in deeper water. Two years later, the highest densities of Nile perch in Kenya were in deeper water, while the haplochromines were completely gone in those areas. In shallow areas, however, low numbers of haplochromines were still caught.

It had been thought that because the Nile perch feeds predominantly on haplochromines, when the haplochromines decreased, the Nile perch would also decline. "Thus, it would result in an equilibrium, so the haplochromines would not completely disappear," Frans explains.

That didn't happen. Nile perch, it turned out, are extremely versatile.

"At the time that haplochromine catches were not more than sixty kilograms [125 pounds] an hour, where before we used to have a ton, the stomachs of the Nile perch were still predominantly filled with haplochromines," Frans says. "Only after the haplochromines had completely disappeared from the Nile perch habitat did this fish start to feed on its own young, on *Dagaa,* the small cyprinid, and on prawns, all of which had increased. This revealed to me that our ecological knowledge of these great systems is poor. Nobody could ever have predicted that the prawn *Caridina* would become an important animal in the lake."

The *Caridina nilotica* he refers to are small freshwater prawns about two centimeters (three-fourths inch) long, detritivores that consume microscopic organic debris. They, in turn, had been eaten by specialized haplochromines. Before the explosion of the Nile perch population, *Caridina* had been fairly rare in the areas of the lake near shore; only occasionally was an individual observed in the catches. But in 1992, when Frans made a trawl at one of the transect stations, he brought up twelve liters (three gallons) of shrimp—about a hundred thousand individuals.

The researchers were also astonished at the changes they began to see in the water of the lake itself. The water had always been somewhat murky with phytoplankton, but now it became so cloudy that it was hard to see anything. With the disappearance of large numbers of fish, especially those which fed on phytoplankton and detritus, organic material decayed and, consuming oxygen, sank to the bottom. Frans's colleagues on the northern end of the lake found that the floor of the lake had been depleted of oxygen: it became an uninhabitable desert.

"Water plants, like *Nymphaea*, that I had not seen in such densities in the shallow areas of the Mwanza Gulf, started to emerge in great densities," Frans recalls. "And then, just by accident at the same time, *Eichhornia crassipes*, the water hyacinth, came into the lake." The water hyacinth—already a problem in parts of the United States —has since become a treacherous weed in Lake Victoria, clogging entire portions of the shoreline, suffocating the littoral fauna and hindering shipping and access to the water by people living along the lake.

Each time Frans returned to Mwanza he would see different and surprising changes, explosions of one sort or another. One time he would find *Dagaa* in the bottom trawl. The next time they would bring up only shrimp; then it would be snails or earthworms. Lake flies also proliferated, in clouds that sometimes forced him and his teammates to stop work. The haplochromines that had once eaten the larvae of the flies were gone.

It took a while to realize that the whole lake ecosystem had been disrupted. "We saw the decline of the haplochromines, and it was still very hard to believe that they would not be there anymore, that there was a part falling out," he reflects. "If I told you that there wouldn't be sparrows anymore tomorrow, you wouldn't believe me, would you? We had problems ourselves believing what was happening before our own eyes. After we realized what was going on, Kees Barel and I informed IUCN [the World Conservation Union] and the press, and it got spread around. There were people who reacted by saying, 'Nonsense—it is not true.'"

He takes out a booklet published in 1986 at the University of Leiden in which newspaper, magazine, and journal clippings are reproduced. In clippings from Dutch-, German-, French-, and English-language publications, the headlines leap out: "Experiment Imperils Lake Victoria"; "African Experiment with Giant Fish Goes Awry, Poses Ecological Disaster"; "Lake Victoria Cichlids Face Extinction";

"Lake Life Destroyed by Cannibal Fish in Food Blunder." Others, just as prominent, tell a different story: "Lake Victoria Tragedy—A Differing View"; "Mbuta Not to Blame for Fish Depletion"; "Nothing Fishy About Nile Perch Plan"; "Nile Perch a Success Story in Lake Victoria." Frans also has a collection of clippings. One, from *New Scientist* (July 21, 1988) carries the headline "Monster Fish May Be Innocent of Ecological Crimes," and the accompanying photo shows two African children with a giant Nile perch. The photo caption reads: "Eating children is one of the few crimes the voracious Nile perch has not been accused of."

The reason for the conflicting headlines in the press, Frans says, was that people could still find haplochromines near shore in parts of Lake Victoria. "They did not know the species, nor what had been there. They came and they still saw many haplochromines, so they believed nothing had happened, nothing was wrong."

The 1986 booklet also includes the text of a resolution passed the previous year at the Fifth Congress of European Ichthyologists in Stockholm. In their statement, the scientists focused on the urgent need to take effective steps to stop further extermination of the native fish in Lake Victoria by the Nile perch:

"Recognizing that never before, man in a single ill-advised step placed so many vertebrate species simultaneously at serious risk of extinction and also, in so doing, threatened a food resource and traditional way of life of riparian dwellers, we, of the Cichlid Workshop, resolve to engender support through the Secretary to the Union of European Ichthyologists and by individual endeavor to conserve the fishes of Lake Victoria." It was supported by more than two hundred scientists throughout the world.

Frans is quick to add that since 1990, some species have indeed recovered in Lake Victoria, but from only two or three of the trophic groups—the zooplanktivores, the detritivores, and the insectivores. Fifteen trophic groups had existed in the lake. He wonders aloud whether the parasite-eating fish—the so-called "cleaner fish"—will ever make a comeback, or the scale-scrapers, or any of the pedophages, which feed on the eggs and juveniles of mouth-brooding females. Or *Haplochromis barbarae*, an egg-snatcher specialized just on the spawning grounds. The list goes on and on. The waters of Lake Victoria had held a world of amazing specialization.

He also wonders why, "on the transect which we monitored since 1977, and of which I thought I knew all the species, there appeared new ones, fishes that I had never seen before." Between 1991 and 1995, seventeen species of haplochromines were collected on the transect across the Mwanza Gulf, and five of these were previously unknown. At the same time, fifteen more species disappeared from the HEST sampling sites near shore.

No one understands the appearance of the new species, but several possibilities are under consideration: for example, old species could be evolving incredibly rapidly in response to the Nile perch, or some very rare species may manage to coexist better with the Nile perch than they did during the reign of the haplochromines, or the "new" species could have come in from other areas.

Frans describes the economic revolution that has taken place on the shores of Lake Victoria since the upsurge of the Nile perch. "Formerly, you had small, artisanal fishermen going around, village fishermen with canoes and with gillnets. Then, when the Nile perch came up, they changed to a Nile perch fishery, but also great enterprises came in, people with money who bought huge beach seines of one-kilometer length, while normally fishermen used a beach seine of two-hundred- or three-hundred-meter length." Local fishermen, whose traditional gear was often damaged by the heavy fish, had to learn how to handle huge new nets and huge new fish.

First in Kenya, then in Tanzania, factories sprang up. In Tanzania alone, the factories could process two hundred tons of fish per day. Nile perch were filleted, chilled or frozen, then flown to cities in Africa, Europe, and the Middle East. The success of the filleting factories pushed up prices for Nile perch along the shores of Lake Victoria, making it very expensive for local people.

Called "elephants of the water" in parts of Africa, Nile perch are many times larger than what the locals had been used to eating. Because most locals cannot afford a whole fish, they now have to buy pieces, and because it is no longer possible to examine the gills or eyes, it is difficult to tell how fresh the fish is. Moreover, the preparation is different, particularly because the larger Nile perch are so fat that they cannot be dried, but they can be fried in their own fat and then held for two or three days.

Most locals still preferred to eat tilapia and other indigenous table fish, but when new means of preservation and cooking were developed, Nile perch became widely accepted. Also rising in popularity is the Nile tilapia, *Oreochromis niloticus,* which was introduced into the lake in the 1950s and has replaced the indigenous tilapiines as a marketable table fish from Lake Victoria. The Nile tilapia appears to be able to coexist without problem with the Nile perch.

Frans has seen the opinion of the local people change several times. "First they said, 'We don't like Nile perch, it's too fat, we can't process it, we don't like the taste,' and so on," he says. "Then when the economic boom came and when it was sold at the local market, people said, 'It's our savior'—this was literally written in newspapers. And then, because people were used to tilapia, catfish and so on—before the Nile perch upsurge there had been more than ten table fishes in the lake, which had virtually disappeared—people said, 'We like some variety, we can only get Nile perch nowadays.'" Recently it has been noted that the Nile perch appear to be changing: they are shrinking, becoming smaller and less fat. Just how much smaller is a point of some debate. But, Frans points out, they are still large enough to eat haplochromines.

Frans feels a deep loss. "I can't say we discovered the haplochromines, because they must have been fished by the local people for a long time, but for more than twenty years we studied their ecology and discovered many species new to science," he says. "In that way, they became part of ourselves. And suddenly we realized we had lost that part."

If there is one plea he would make, it would be that the scientific community re-evaluate the need for what he calls alpha taxonomy and species inventories, which have fallen out of favor in the face of emphasis on experimental work. Even though the reason he himself went to Africa was to collect data for experimental work, he acknowledges: "If we hadn't done the inventory, we would have missed a great event and nobody would have realized how serious it was."

He believes that scientists have a kind of a moral obligation to record what they find in nature. "You can't turn your back to it and say, okay, my plans are running in another direction and that's the way I am going on, it's not my business. It *was* our business, because we happened to be there."

The drama of what happened in Lake Victoria, he believes, should not be minimized. "Some people say it's a nice, great experiment . . . and then indulge in the natural experiment that's going on. But I think that experiment should have never happened."

The astonishing variety of life on earth brings human beings esthetic pleasure, he continues. And now that the differences are being lost, the result threatens to be monotony. A certain level of diversity is also important to keep ecosystems functioning, although "a lot of people may not believe it."

That Lake Victoria has dramatically fewer fish species is already clear. Whether the total weight of fish in the lake will also be lower at equilibrium—if a new equilibrium is ever reached—is not yet clear. "We can make all kinds of nice models, ecological models, but I am afraid that we are still too stupid, at least have not enough data on this lake, to make real predictions. Nobody ever predicted that the shrimp would play such an important role, but it is playing it now. And I don't know what happens next."

What sort of knowledge may have been lost with the extinctions of the hundreds of species from Lake Victoria?

"We had the opportunity to look into the kitchen of evolution here, because the system together with the systems of the other two lakes, Malawi and Tanganyika, gives just such a nice cross-section to what happened over the past seven million years in this area with those cichlid fish." Lake Victoria was believed to have developed between 750,000 and a million years ago, a very short time compared with Malawi, which is a few million years old, and Tanganyika, which is perhaps seven million years old. Recent borings in the bottom of Lake Victoria revealed that a large part of the lake, possibly the whole lake, dried up as little as 12,400 years ago. This estimate makes the rate of speciation that produced the diversity of fish in the lake even more spectacular.

"We've lost the young lake where apparently things were still ongoing," Frans says. "We started to unravel some ecological and evolutionary questions, but the fishes that we studied are no longer there." The number of individual haplochromine fish, at least those belonging to the species that are left in the sublittoral and deep-water areas, has been reduced by a factor of a thousand. Frans explains that hybridization among the species is therefore to be expected—not only because females are desperate to find males, but because they make "mistakes" in the murky waters and cannot always recognize the males of their own species by coloration.

He excitedly describes experiments done in Leiden that show how the probability of hybridization is related to water transparency. In turbid water, certain colors are no longer transmitted and a fish ready to spawn may accept a mate whose inappropriateness is difficult to perceive in the murky water. In a 1997 paper in *Science*, he and Leiden colleagues Ole Seehausen and Jacques van Alphen reported laboratory experiments showing that turbid water curbs sexual selection, increases hybridization (a process in which two pre-existing species merge to form a single hybrid), and thus furthers the decline in cichlid diversity.

Frans is enthusiastic about the results of the experiments. "That's a nice thing," he begins—then stops, embarrassed. "Oh, nice thing— there I go, you see, there I go. I'm saying it is a nice thing. No, it's an awful thing." He is clearly torn between his scientific curiosity and the damage that has been done to "an incredible system."

"Just to pick out a few of these things, it would have been much better to have studied the system *in situ* and find out what the original system was."

It is crucial that the data and preserved material, including many nicknamed species and unsorted samples, remain to be studied and properly described. "Only such data can show what really happened and still is happening in Lake Victoria." Extinction of species can, of course, always be questioned if the species were never documented. "There have been a number of people saying it's nonsense what they [say] about this extinction, it's not true, there're not so many species. Unless the species are described and we really have circumscribed these entities, the diversity is difficult to believe."

Something else that he has noticed he finds both interesting and disturbing: "People don't believe in species until they are described." He points to the aquarium trade, where this is frequently seen. "As soon as the fish have a scientific-sounding name and are described or pictured in books, people are more eager to buy them." He has also known scientists who were reluctant to study unnamed species.

Frans and his colleagues found, altogether, about five hundred species of haplochromines in Lake Victoria, half of which he considers now extinct. It is difficult to know how many existed in actuality, particularly in the deep water of the lake, which still has not been thoroughly investigated. It may have been a thousand species, he says—no one knows. What he can say with certainty is that within a decade, perhaps just from 1982 to 1987, nearly half of the species found up until then had disappeared.

He cautions about the arrogance of making judgments about the people who introduced the Nile perch. He relates a story from the time he was in Mwanza, Tanzania, when many trees were being cut down. He asked the people who were cutting down the trees if they realized that their children would not have anything to cut, and that the area would become a desert. "And they said, 'Yes, we realize it, but if we don't cut them, we will not have something to make our food with tonight, and we will die now.' And that was for me quite shocking. . . . We are in a luxurious position with all kinds of alternatives, and we first damaged a lot, and now we are worrying about nature. They are in the situation that if they do not cut the trees now, their children will starve—they cannot think of the future of their children, even."

But although Frans understands that each generation needs time to learn, he looks at the enormous population of today's world, mostly people younger than thirty-five, and realizes that if they all have to learn by making the same mistakes that earlier generations have, little will be left to learn from or to save.

Kees Barel is a functional morphologist—that is, a biologist who works to explain the adaption of anatomical structures. To obtain data on the form and structure of cichlid fish, he has had to dissect them and analyze the mechanical properties of muscles and bones that enable the fish to survive and reproduce in their environment. He has spent considerably more time in the laboratory than in the field. By the time he has completed a dissection, part of the fish will have been destroyed to gain needed information. Following a carefully designed anatomical procedure, he tries to avoid unnecessary damage to the fish and to minimize the number of specimens needed to fuel his research over the years. A thoughtful man, he has reflected even more carefully in recent years before taking a jar off the shelf of his laboratory at the University of Leiden to dissect a haplochromine from Lake Victoria. These preserved fish have come to represent a dilemma for the morphologist, who aspires to know more about the structure of the fish and yet realizes that by cutting into one of them, he may destroy the last whole specimen in existence.

"There came a sad moment when I realized that most of these species are extinct," he says. "It gave me the feeling that dissecting these fish is like cutting up a Rembrandt to learn what paints he used." No words are available either in his native Dutch or in English, he says, to describe the depth of his sadness at their disappearance.

"My first field team caught them by the score," he remembers, "and then within a few years I had to accept from the team's reports that they were no longer there. Even after fifteen years I cannot come to terms with this loss." Some of the specimens he is working with come from his first foray into Lake Victoria in 1975.

His analogy of fish to art is apt. Growing up in The Hague, a city famous for its museums, Kees developed an appreciation for both art and literature. In fact, he says he would have studied literature, but he didn't want to spoil it with "academic analysis." He has never regretted going with his second choice: biology. He studied fish mechanics and, inspired by his supervisor, Jan Osse, decided to do his thesis in the field.

"My main interest in biology was, and still is, anatomical variation on a theme," he explains. "I preferred comparative functional morphology to developing biomechanics *per se* and testing a hypothesis experimentally on a single, technically suitable species."

On nearing the completion of his doctorate, he wrote to Humphry Greenwood to seek advice about what group of fish to work on next. When to his astonishment he received a reply from the renowned scientist within a week, Kees traveled to London to talk with Greenwood. Thus began his friendship with the man he credits with being "the father of our research."

Their meeting was memorable: "This world-famous ichthyologist received me in his shirtsleeves! His room was bare, smaller than mine, and his work table at the window was full of greasy, labeled jars containing a urine-yellow spirit with boringly similar fish. He advised me to study cichlids and pointed to the jars."

Most cichlid species are found in the Great Lakes of Africa: Malawi, Victoria, and Tanganyika. Kees says he would not be surprised to learn that each of the lakes once contained a thousand species of fish at the time. Following Greenwood's advice, Kees started research in 1972 at the University of Leiden, not on the diverse cichlids of Lake Victoria, which Greenwood knew were difficult to distinguish, but on sixteen haplochromine species from the much smaller Lake George, in Uganda.

Thus, he already had experience in Africa by the time Gerrit Anker saw that fateful television program about Prince Claus on Lake Victoria which diverted the focus of their research to another lake, with a profusion of species: "We became involved in the cichlid fish of Lake Victoria, which would dominate our research for more than twenty years and which would make us eye-witnesses to the greatest mass extinction of vertebrate species in modern times."

Besides advising Kees to do comparative research on cichlids, Greenwood had also warned him at that first meeting in London that "science is a lonely affair." Kees was surprised to hear this from a man of such prominence in the scientific community. Later on, however, he came to understand that systematics is usually so specialized "that it is rare to deal with a systematic subject where there is another specialist with whom you can discuss [it] with the same level of expertise." As the Leiden group struggled with the practical problems of identifying the Lake Victoria cichlids, Greenwood helped ward off loneliness by lending his expertise. He traveled to Leiden to talk with the group about the species that he knew so well and on which they also would become experts.

Kees reminisces: "These were moments of esoteric pleasure, to open a jar, take the fish out, hear Greenwood say, 'Yes that is good old *parvidens*' and not only knowing what he meant by *parvidens*, but also by *good* and *old*." It was one of his greatest experiences in biology, Kees says, to discover how difficult taxonomy is. "It is not simply 'A rose is a rose is a rose.' Ask a rose specialist what a rose species is and he will tell you in essence about all the [same] problems we had and still have in distinguishing cichlid species from Lake Victoria."

The loneliness of the work has now acquired a broader dimension than that of of working on the systematics of little-known fish. Not only have many of those fish disappeared since Greenwood started working on them nearly half a century ago, but the scientists who worked on them are also passing from the scene. Kees lists their names, from F. M. Hilgendorf, who described the first cichlid from

Lake Victoria in 1888, to Ethelwynn Trewavas, who died in 1994, and most recently Greenwood himself, who died unexpectedly in 1995. "It is now only Frans and Els Witte, and Martien van Oijen, who know what I mean by the excitement of sorting out trawl catches from the pristine species flock," Kees says. "Our Ph.D. students have never seen them and it is difficult to convey your feelings. They say, 'Why do you talk about lost species so much? There are still so many left.' That's true, but their diversity is an impoverished relic of what I have seen with my own eyes, when I was there only twenty years ago."

This is why, when he lectures now, he shows pictures of catches as they were before and after the Nile perch boom. "I say, look here, these species you see *no longer exist*. There were such exceptional things as a scale-scraper and a parasite eater and pedophages among the Lake Victoria cichlids.

"I can derive from anatomy the same esthetic pleasure as from a good painting," he continues with careful emphasis. He describes how, through his microscope, he can journey into the world inside a cichlid's head, to observe, within a few cubic centimeters, "the sheer miracle of how crowded muscles, nerves, and blood vessels are packed." To unravel how these fish, exhibiting only minor variations from one to the next, managed to exploit nearly all resources in the lake is the greatest challenge of his research, he says. "For that reason, I selected them, and now they are gone." Gone, too, he adds, is the natural environment necessary to understand the adaptive nature of the small anatomical differences of such similar fish.

He has been much affected by what he has witnessed in Lake Victoria. He reflects that in the early days of work on the lake, the researchers had no quick means of communication, and it was hard for the team to convey findings about the species to colleagues in Leiden. They kept logs or diaries that ran to hundreds of pages and described not only the species they found but the emotions of the scientists as well. Rereading those diaries now, with their comments and exclamations in the margins of the watermarked pages, he keenly realizes that "those were great moments, when those discoveries were made." He recalls the nearly unbelievable reports in the early logs on the discoveries of more and more new species; then when the first transport of preserved specimens arrived in Leiden, it more than confirmed the reports. It was, he says, "like birthday presents when one is young," and describes the scene as a huge box was opened, revealing an incredible array of new species wrapped in cloth.

The realization that the haplochromines in Lake Victoria were going extinct came piecemeal, he remembers, but it came rapidly. There had been rumors from Kenya in the early 1980s that something was wrong. Then, as Barrel and Witte summarized the data in Leiden, they could not ignore the conclusion that many species had disappeared. It was then, in 1985, just a decade after their discovery of the incredible wealth of cichlid species in Lake Victoria, that Kees and Frans traveled together to Gland, Switzerland, to the headquarters of the World Wildlife Fund and the World Conservation Union to announce the extinctions.

One can easily imagine, Kees says, what might have happened had no detailed data or collections been assembled on the taxonomy and ecology of the lake's fish and other fauna, in particular the cichlids. Their claims that a series of extinctions was under way would have been challenged; the researchers would have confronted assertions that with so many species left, not many could have disappeared. The question was not, Kees points out, whether any haplochromines were left; "The question was what species number and ecotypes of haplochromines there were prior to the Nile perch boom, compared to what is left."

Knowing many people who have worked in tropical systems, he realizes that it is rare for anyone to know with certainty what species were present before a decline. "But in our case, we could demonstrate it," he says. This is why the fish in the jars in the museum are so important. Only with these as evidence can one say what has become extinct in nature. That is also why Kees had been eager from the beginning that there should be a museum collection—although he admits he thought that it would be the development of fisheries on the lake, not another fish, that would destroy species on Lake Victoria.

He gives credit to Frans Witte and Martien van Oijen for keeping the extensive "field books" of fishes caught that contained all their notations. It is predominantly on the basis of the field books that any quantitative estimate of the loss of species might be presented. Scientists in Leiden are still extracting information from the books.

What would the world have missed if the Leiden team hadn't been there?

"Diversity first of all," he answers without hesitation. "These fishes are a unique case of diversity. There is no other group of vertebrates with so many species and such an ecological diversity realized with so little anatomical modification."

The Leiden team had monitored the lake's ecosystem before the Nile perch upsurge as well as during it, so the background of the changes was known. "If you had not known there were phytoplankton-eating cichlids and snail- and chaoborus-eating cichlids," Kees remarks, "you could never explain why all of a sudden there was this eutrophication of the lake, this enormous bloom of chaoborus larvae or this enormous production of snails. The dramatic effects the Nile perch had on the whole ecosystem were *unpredictable*. Nobody could have foreseen it would become a fish fauna dominated by only three species, that there would be an enormous increase in shrimp, and that the Nile perch, to survive, would eat its own young, which in turn feed on zooplankton."

Kees has accepted many invitations to speak about what has happened in Lake Victoria: from schoolchildren, universities, travel clubs, and gatherings of aquarists. He considers these talks as important as lecturing at scientific meetings. He also talks to his own children. "I want them to remember that something beautiful, diverse, and complicated, in which I was emotionally involved, has been destroyed. That's a feeling I can never pass. This is a catastrophe in my life—I can't compare it with going through a destructive war, but in my life, the decimation of a species flock from Lake Victoria is a tragedy for which I still can't find the appropriate words."

Lake Victoria is not a small lake, he emphasizes. "A large ecosystem has been destroyed in Africa." Compare it, he suggests, with Africa's Serengeti, with all the large animals gone. "Twenty-five years after my first meeting with Greenwood I have come full circle," he concludes. "All I then had available of the Lake Victoria cichlids was preserved museum specimens and, of most species, that is all that is left now." The difference, he notes wryly, is that he can no longer imagine why, at that meeting with Greenwood, he found the fish so boringly similar.

Jinja, Uganda, an hour's drive east from Kampala along the northern shore of Lake Victoria, may be best known as the source of the Nile, which flows northward from here through Lake Kyoga on its

Lates niloticus. Ancient Egyptian representation of two fishermen carrying a Nile perch. This drawing was taken from the grave of R'htp Medum (ca. 2500 B.C.). From the original color print in W. M. Flinders Petrie's excavation report "Medum," London, 1892.

way toward Sudan and Egypt. Jinja is also the location of Uganda's Fisheries Research Institute (FIRI), a cluster of modest low buildings just a short walk from the pier where the research vessel, *Ibis,* is moored. From here, African scientists have watched, through war and peace, poverty and prosperity, as the consequences of introducing the Nile perch have unfolded.

Sylvester Wandera, a research officer at the institute, did his undergraduate work at Makerere University in Kampala and went to Dar es Salaam, Tanzania, to earn a master's degree in fisheries and aquatic science. In the course of his studies, in the early 1980s, he worked on haplochromines with HEST in Mwanza. One curious trophic group, the pedophages, fascinated Sylvester because of their specialized feeding habits. They "engulfed" the snouts of female fish that had taken eggs and hatchlings into their mouths to protect and brood. Pedophages would suck out the eggs and young offspring and eat them. Their other technique was "head-drumming" a female in order to make her spit out the eggs or hatchlings. In studying four engulfer species, Sylvester was intrigued both by their diet of mostly eggs and hatchlings and by the way their morphology had evolved so that they could feed effectively. "Evolution didn't take one line," he says, and explains how the pedophages' teeth adapted to become not very sharp, deeply embedded in the oral mucosa, and capable of biting over the parent's mouth. Although he only worked on four engulfer species, he says that ten or more may have lived in the lake. Sylvester last saw them in 1984. They have probably been destroyed by Nile perch.

Plenty of haplochromines were left near Jinja when he went to study in Dar es Salaam, Sylvester says, although Nile perch were not uncommon. He returned to find a greatly increased number of Nile perch, and fewer haplochromines. Then, in 1984, the research vessel *Ibis* broke down, and for three years Sylvester and other Jinja scientists could not go out on the lake to take samples. When they did return to the lake on another, smaller trawler in 1988, most of the haplochromines were gone.

"I had just such a developing interest in them, and when they disappeared, I felt cheated," he says. Wondering aloud if they could all have continued to coexist, he reflects that he probably thinks about this only because he is a scientist. "The ordinary people, of course, feel happy that there is now more fish to eat," he says. "There is good business, and the economic activity on the lake there is booming, it has been booming for quite some time just because of the Nile perch in the lake. There is also this thing about the haplochromines—after all, they would say, nobody was eating them." He laughs at the irony of his situation and adds: "They were only of interest to science."

What exactly was their significance to science, and do the economic benefits of the Nile perch offset the haplochromines' loss?

"You see," he explains carefully to his visitors, "for a long time the haplochromines had been used as examples of adaptive radiation and evolution—and these present very good examples, something you see. For us Africans here, you will talk of Darwin's finches, but we do not know them. But surely, when you talk about haplochromines, we understand what you are saying. You look at them and see the difference, this one here, that one there. To the layman they look the same, but when we start looking at them in detail, they become very interesting, and they were our local examples of the subject. So economics,

I cannot talk of it, just to say they were not useful. But fortunately, we have a few, overlooked, still somewhere here."

He says he doesn't believe anyone could have predicted what would happen when the Nile perch were introduced into the lake. It was expected that the Nile perch would eat other fish, but no one expected the perch to devour them at the rate they did. People "never once figured that they would eat their food in four years and finish them."

According to Sylvester, most of the open-water species of haplochromines in Lake Victoria are gone. Other species of haplochromines have survived along the edges of the lake. Those which survived are species originally restricted to habitats difficult for Nile perch to penetrate, such as rock crevices and vegetation along the shore. He is reluctant to estimate how many species have become extinct. "Very many," he ventures—and when pressed: "More than fifty."

Some of the haplochromines that disappeared, Sylvester believes, were bottom-dwelling species that went out not only because of the introduction of the Nile perch, but also because of the change in conditions on the bottom of the lake, the anoxia that has resulted from the eutrophication of the lake.

"The lake is getting old," he says with a sigh. "Many people feel that the current boom of the Nile perch may not be sustainable, because as the lake worsens, the anoxia increases, there is less space now left for fish. I hope it doesn't go that way. But if you really look at it, you certainly fear for the lake. This may just be temporary, but"—he lowers his voice— "the lake may die."

Timothy Twongo, principal research officer at FIRI and leader of the limnology program, has studied the water hyacinth, an invasive plant that contributes to anoxia in Lake Victoria. Timothy had gone from Makerere University to get a doctorate from the University of Guelph in Canada, then to the marine research institute in Zanzibar. In 1978 he returned to the institute in Jinja. He first saw water hyacinth in Africa in May 1988 on Lake Kyoga. Having seen it earlier during a trip to the Philippines, he was concerned when it appeared in Lake Kyoga and warned the fishermen that the weed would become a problem unless it was quickly removed. They didn't believe him. "They were laughing—they didn't understand," he recalls.

Water hyacinth at first grew in isolated mats in the eastern portion of Lake Kyoga, and a few mats floated near where the Nile flowed in from Lake Victoria. By the end of the year, patches of the weed had broken up into little mats and were floating all over. When he did his first survey in 1989, they had spread completely across the lake. Water hyacinth was reported in Lake Victoria that same year.

By 1990, the fishermen on Lake Kyoga no longer laughed, they were already encountering problems: several landings were periodically clogged with hyacinth, making it difficult for them to get to the lake to set their nets or to bring catches ashore.

"By '92, we were getting alarm bells about water hyacinth in Lake Victoria and by '93 it was a definite problem," Timothy recalls. Within four years, water hyacinth mats had become solid in places along the edges of the lake. "Resident mats" that had not broken free, averaging ten to fifteen meters (thirty-five to fifty feet) across, ringed portions of the shoreline. Mobile mats float free and may be blown by the winds into protected bays, where they accumulate in large concentrations. They thrive particularly in waters near city outflows, which are rich in nutrients. Timothy tells of one mat in the bay near Kampala that covered six hundred hectares (fifteen hundred acres).

Resident water hyacinth often grows in association with hippo grass (*Vossia cuspidator*), which makes the mat solid enough to walk across. Its roots may dangle a meter into the water below. In addition to clogging the water space, the water hyacinth markedly reduces the oxygen concentration in the areas of the lake it covers. In the mud at the lake bottom, oxygen levels may be reduced much more. Timothy believes that not only do the fish suffer in such conditions, but their eggs and larvae may die.

The classic means of controlling water hyacinth is simply to remove it. It is also possible to control the plants biologically, but that method, which has been initiated in Lake Kyoga, takes three to five years to begin to work; its effects are now becoming apparent on the mobile mats on Lake Kyoga. The chemical option, which is used in Florida, involves applying herbicides every year. Timothy is not convinced that the chemical option is a feasible means of control over the long term, given its expense. The impact of the chemicals on other aquatic life forms has not been demonstrated.

John Okaronon is the FIRI research officer responsible for estimating fish stocks in Lake Victoria. He cites a survey carried out in the Ugandan part of the lake from 1969 to 1971 from the *Ibis*. In the survey Nile perch were at first listed as making up less than 0.1 percent of the available stocks; haplochromines, about 83 percent. The big change came between 1979 and 1982 in the Ugandan part of the lake, with the commercial catch of Nile perch increasing tenfold while that of haplochromines decreased twentyfold. By 1993 to 1995, the Nile perch accounted for 96.7 percent of the research hauls by weight, compared with haplochromines at 0.2 percent.

From 1982 to 1988, fish hauls in Lake Victoria rose dramatically overall. John points out that although fishing activity has increased substantially in the Ugandan sector of the lake, from thirty-two hundred canoes in 1972 to forty-five hundred in 1988 and to more than eighty-six hundred craft by 1992, yields are now declining. The catch in the lake decreased from 132,000 tons in 1989 to 120,000 tons in 1991 and 103,000 in 1994. Parallel trends have been documented in the Tanzanian sector.

Trained in botany and zoology at Makerere University and in fisheries science at the University of Michigan, John has worked at the institute in Jinja since 1972. Since 1989 he has been the Ugandan leader of the Lake Victoria Regional Fisheries Project, working in coordination with fisheries research officers in Kenya and Tanzania.

John estimates that 60 percent of the species of haplochromines caught twenty-five years ago are extinct now. He puts the number of extinct species at about 180. Even though the haplochromines may not have been sought after by the local people for food, he—like others at FIRI—sees a link between the haplochromines and the health of the lake. By feeding on detritus, the plants and nutrients from agriculture on the lakeshore, the haplochromines represented a natural control against eutrophication. When the haplochromines disappeared, no other fish filled that role.

The Nile perch, John points out, has created a prosperous food industry. When it comes to table fish, "we have more now than we had twenty-five years ago, because twenty-five years ago when we had 800 kilograms [1,800 pounds] per hour as a measure of the abundance, over 80 percent of this was haplochromines, which were not table fish: it was considered trash." Only about 10 percent of the fish caught then, the tilapia, were desirable food. But now, John says, the fishing yield from a trawl net in the lake is 160 kilograms (350 pounds) per hour and more than 90 percent is Nile perch, all table fish. He adds, apologetically: "So that is the way we try to console [ourselves] a bit, but that is at the expense of the environment."

In 1995, John says, seven thousand tons of Nile perch with a value of fifteen million dollars was exported from Uganda as filets; they came from seventy thousand tons of Nile perch, processed in twelve factories employing fifteen hundred people on the Ugandan shore of the lake. The fishermen's income, figured at one thousand Ugandan shillings per kilogram, was seventy million dollars at the landing.

Although the price of Nile perch is rising, the catches seem to be decreasing. But the situation is not clear. John emphasizes that the figures *may* reflect an actual decline in the stocks, or they may reflect how much fish is going to "a neighboring country," which he would not name. He estimates that three hundred metric tons of fish per day crosses unrecorded to a neighboring country—or half of what is caught and recorded in Uganda. Because some Lake Victoria fishermen appear to market their fish in another country in exchange for commodities that they sell when they return home, it is difficult to get reliable catch figures from the three countries that share the lake— and difficult to know whether the lake is being overfished.

John confirms that Nile perch in Lake Victoria are getting smaller. In the 1980s, Nile perch caught in a trawl commonly weighed up to eighty kilos (175 pounds) each. Nile perch caught in the mid-1990s averaged about eight kilos (20 pounds), according to John. In response, fishery officials in the region have agreed on a five-inch (127-millimeter) gillnet mesh. Seine nets, which destroyed fingerlings, are already banned. Trawl nets, if permitted at all, John believes, should be allowed only in waters deeper than sixty-five feet (twenty meters). The key to these regulations, however, is coordination and enforcement in the three countries that surround the lake—no easy task.

Asked if he thinks it makes any difference that 180 haplochromine species have gone extinct, John replies that he has two views. "From the layman's point of view, I think the loss is [compensated] by the importance of the Nile perch," he says, citing the economic benefits, including increased employment opportunities at the fishery.

"From the scientific point of view," he continues, "the Nile perch has caused a disaster."

Lycaon pictus. Drawing of an African wild dog by Lorene Simms.

8

HANDLE WITH CARE: A WILD DOG STORY

Whose animals are they? Whose dogs were they for me and others to research? Why was it that I and others suddenly arrived in the Serengeti and did what we did to those animals? It appeared to me that it was a very haphazard event.—ROGER BURROWS

You can't say, just because it's going to create a bit of a stink, we shouldn't say handling wild dogs is bad. If that means I can't handle hyenas, so be it. . . . I can do my research and everybody in this place can do their research without collars.—MARION EAST

 IN THE SUMMER OF 1996, five years after the disappearance of all African wild dog study packs from the Serengeti National Park in Tanzania, posters at Naabi gate and in park lodges still appealed to visitors: "Wild Dogs Need Your Help." Under a large photograph of a young female wild dog (*Lycaon pictus*) and a puppy, the text explains that the wild dog is the rarest and most endangered of the large carnivores in the park, and that the known adult population numbers thirty-one. "If you see a wild dog you are very lucky indeed, not only because they are so rare but because they are nomadic," it states. It describes a project under way in the park to monitor the study packs and urges visitors to report wild dog sightings.

All the wild dog packs being studied in the Serengeti disappeared in 1991. They did not simply become rarer, they became extinct. If you are lucky enough to see a wild dog in the former study areas now, it is a stray from somewhere else.

Why do these posters continue to be prominently displayed, leading visitors to believe that the park still has a wild dog population?

If an animal is one of the most endangered mammals in Africa and lives in what scientists assert is one of the best understood ecosystems in the world, isn't it carefully monitored, and wouldn't there be an immediate and loud outcry if it died out? Wouldn't the public be informed? How did the dogs die and why? What part did disease play in their demise—and what role did human beings play?

The questions are apt; the answers remain controversial. Sometimes the facts are unknown, sometimes difficult to discern, perhaps sometimes obscured. The dogs themselves are difficult to follow, simply because they are nomadic. Individuals move from one pack to another; packs move from one geographic location to another and range across borders and over huge distances. Now it appears they have entirely run out of space in the Serengeti. Their story illustrates the haphazard way in which some wild animals—endangered or not—are studied, handled, or "protected" by methods that don't help and may even harm them.

Not only did scientists not inform the public; they said little to each other in 1991 and 1992 about the disappearance of the wild dog study packs. At a conference of African wild dog experts in Arusha, Tanzania, about 225 kilometers (140 hundred miles) from the Serengeti, six months after the collars of the last dogs in Serengeti study packs had been found, the disappearance was barely mentioned. In fact, the man in charge of the Wild Dog Project in the park, Markus Borner, who represents the Frankfurt Zoological Society, maintained before the audience that he was still monitoring forty-two dogs. A month later, a hundred-dollar reward for information leading to the finding and photographing of wild dogs was posted in the park and flyers were sent to tour companies in Arusha. No one ever collected the reward.

Among those who knew the dogs were gone, however, were three scientists: Roger Burrows, who had most recently studied the Serengeti dogs, and Heribert Hofer and Marion East, Max Planck Institute researchers who worked on hyenas and were at the Arusha conference as "observers," not participants. All three were deeply disturbed by the Serengeti extinction, by the way it was quietly accepted or covered up. This story is written from their viewpoints.

For other viewpoints, many of which were expressed in scientific papers after the fact of extinction could no longer be ignored, see the bibliography.

Roger, who put up the wild dog posters in the park in 1990, was the first scientist since the 1970s to receive research clearance from the Tanzanian government to study the African wild dogs. He had examined the records and knew that in the years 1986, 1988, and 1989, each time a study pack had disappeared and empty radio collars from individuals in the pack had been found, the pack was never seen again. On the plains of the Serengeti, where nothing goes to waste, even bones disappear into the jaws of hyenas and jackals or are carried away by vultures. The empty radio collars that scientists use to track the wild dogs are usually the only evidence of the animals' death.

In late September of 1991, after empty collars from the last three study packs were found, Roger left the Serengeti, convinced that all the study packs were dead. He was astonished to hear that his posters were still being used in the park five years later. But after giving it some thought, he observed, "It doesn't surprise me in some ways, in the sense that certain people in the Serengeti never admitted that the dogs had disappeared at all—there was an attempt to completely cover the losses."

His story of the two years he spent in the park may surprise any conservation-minded citizen who visualizes well-coordinated, well-funded efforts to protect and study endangered species.

In 1989, Roger decided to take early retirement from the University of Exeter in England to do research in Madagascar and thus fulfill a dream of many years. After talking with scientists at Kew Gardens in London about their needs, he settled on a project to photograph pollinating insects on the plants of Madagascar. He describes himself as an old-fashioned naturalist, interested since childhood in the "tremendous diversity of animals and plants and the way they integrate together." He recounts his asthmatic childhood, how he had spent long periods lying in bed reading natural history books and looking at the trees through the window, longing to be "out there." And that is where he says he has always been most at home, in the outside world, in nature. "Man's bits and pieces, the buildings and all, I love them, but that's an artificial world that we create."

He earned a degree in biology and geology from the University of

Keele, taught in secondary schools in Shropshire, Derby, and Cornwall, and was at the University of Bristol doing doctoral work on red foxes when his fieldwork was cut short by a serious car accident. After his convalescence, he developed an interest in marine biology and, until financial pressures forced him to close it, ran the first independent field station in the United Kingdom. The timing for the opening of the field station was unfortunate: just as the first students arrived to do research at the field station on the pristine beach Roger had chosen in Cornwall, the *Torrey Canyon* ran aground offshore and spewed thick black oil all over the coastline. He switched gears quickly and rallied the students to clean oil-soaked seabirds and later to record the effects of oil on the marine life of the rocky coast. Although most of the oil-soaked birds died, he was happily surprised to see the major visible effects of the spill disappear in about two years. He ran the field station for nearly a decade and remembers those years as some of his best.

He then went to Exeter in 1976 to work in adult education. While there he organized and led safaris that were based at Lake Naivasha in Kenya and also explored the Masai Mara Game Reserve. When he left Exeter in 1989, with visa for Madagascar in hand, he led one last safari in Kenya and put the tour participants safely on a plane back to the United Kingdom. Then, with an invitation from friends in the Serengeti, he crossed the border into Tanzania to make a short visit to the famous wildlife park before heading off to Madagascar. The friends, wildlife filmmakers Mark Deeble and Victoria Stone, were camped on the Grumeti River, home of some of the world's largest crocodiles. Roger marveled at the Serengeti's vast expanses and remarkable wildlife and was introduced by Mark and Vicky to some of the scientists working in the park. Toward the end of his stay, Roger recounts, "the question came up whether I could help do this wild dog project which was under way. I'd never seen a wild dog and I'd only seen pictures, barely knew what they were." But after talking with Markus Borner and Karen Laurenson, a scientist working on a cheetah project who had been advising Steven Lelo, the Tanzanian wild dog project officer, Roger decided to extend his stay in the Serengeti. He became a senior researcher and adviser to Lelo, who had joined the team six months earlier after working as a tour guide at

Ngorongoro Crater. Roger's trip to Madagascar was postponed, never to be realized.

Over the next two years, Roger became the only person in the park who could identify individual wild dogs. His long hours of observation enabled him to describe and understand the dogs' demography and behavior—but in the end he would remain bewildered by the behavior of some of his fellow scientists.

Even though he never foresaw a long-term project, he applied for research clearances from the government and the parks department. On the research applications, at Borner's suggestion, he listed the Frankfurt Zoological Society (FZS) as his sponsor. The Serengeti Ecological Monitoring Program (SEMP), set up in 1986 for a three-year period and funded by the WWF and the FZS, had monitored cheetahs and wild dogs and had provided money for other animal counts. In September 1989, major funding of SEMP via WWF ceased. The Hunting Dog Program became the Wild Dog Project, which continued with funds from U.S. sponsors, Neil and Joyce Silverman, administered by the FZS, which provided logistical support.

Roger, however, was financially on his own, except for a petrol allowance from the FZS, which he received later in the project. He drew no salary and learned the ropes—and dogs—as he went. He also provided his own transport. Borner promised that repairs to any vehicle Roger could get as far as the Serengeti would be made by mechanics at the FZS base in Seronera.

It was necessary for Roger to leave Tanzania and then return once official research clearance from the Tanzanian government's Commission for Science and Technology (COSTECH) and permission from Tanzanian National Parks (TANAPA) had been granted; both were necessary for work in the Serengeti.

So, in October 1989, Roger left Tanzania and went to the National Museum in Nairobi, where he sought out Pieter Kat, the leader of a wild dog project who was studying packs in and around the Masai Mara Game Reserve, the main extension of the Serengeti ecosystem in Kenya. "I knew that there had been little contact between the Tanzanian wild dog project and the Mara project," he explains. Roger thought it essential to familiarize himself with the whole ecosystem and its wild dogs and to get in touch with Kat, possibly with a view to adopting similar techniques and policies. He came away impressed by Kat's well-organized system of cards with color photographs that could be used to identify every wild dog in the study pack.

At the Nairobi meeting, Kat urged Burrows to recommend to Borner that all Serengeti packs be vaccinated against rabies, a policy that Kat had recently adopted in Kenya. He did not tell Roger then that some members of the Aitong pack, the only regularly monitored study pack in the Mara area, had been vaccinated against rabies beginning in 1987. This pack, including two individuals vaccinated in 1989, had all died between August and September 1989, just prior to Roger's visit; the presence of rabies had been confirmed in some of the dead dogs.

Roger had not been aware that Serengeti wild dogs were threatened with rabies, and he was surprised that Kat was so insistent on the need to vaccinate all wild dog packs. Nevertheless, he assumed that the vaccinations were safe and effective, that the techniques to be used on this endangered species had been fully investigated and were based on established veterinary experience and procedures. All these assumptions proved wrong.

After arranging to collaborate with Kat whenever possible and to give him the radio-collar frequencies of the Serengeti dogs so that Kat could monitor immigrants from the south, Roger flew to Dar es Salaam to get the necessary permits. On returning to Arusha with research and residence permits, he set off for the Serengeti with Jan Corlett, friend-*cum*-research assistant. They settled into housing rented from the Serengeti Wildlife Research Centre in the park. At first Roger borrowed a vehicle, but he soon returned to Arusha and bought a battered Land Rover van, formerly the property of the British army, which was to be their home during long observation periods. It was, in Roger's words, "ancient when it went to Africa, even more ancient after being used on those roads."

Markus Borner suggested to Roger that his first order of business should be to locate all the dog packs and groups in the Serengeti and adjacent area and, with Steven Lelo, to radio-collar them. No one knew how many dogs there were or where their territories were located.

In late 1989, as Roger started to work, the data he had available consisted of a report written by John Fanshawe in 1988, mainly on the Naabi pack, a list of chance sightings of dogs by tourists and others, and the location and date on which study packs with collared dogs had been radio-tracked from the air. Fanshawe and his colleagues had also reported on two other packs: the Pedallers, out of the park in the Ngorongoro Conservation Area, and the Ndoha pack in the park's western corridor. The Pedallers had died out in June 1986. Rabies was suspected as the cause, but it was never confirmed.

Roger soon realized that although many tracking flights had been made, in only a third of them had records been kept of the number of dogs seen from the air. Even rarer were attempts to follow dogs on the ground after they had been located from the air. In most cases, all that was known was that a radio collar had been located at a particular place on a certain date; whether the dogs were alive or dead at the time was often unknown.

During its decline to extinction from May to September 1988, for example, the Naabi pack, one of whose members had been radio-collared in May, was located six times, but because it was not visited on the ground, no tissue samples were collected from the twenty-four dogs that disappeared, and the cause of the death remains unknown. One serum sample had been taken from a collared dog in May, but the label was lost; it is not known whether that individual had been exposed to rabies.

Some photographs of individuals in the study packs were available but had not been updated for two years. Roger estimated that of 140 individuals that could have been photographed, only a third appeared on the cards. Moreover, no written record had been kept of the radio frequencies of the collared dogs or of the dates on which they had been collared.

With such data, Roger began his work.

Steven Lelo had been able for a brief time to study the Ndutu pack, which died of unknown causes in late 1989, as well as the Salei pack formed in 1988, which was frequently seen near Naabi gate. The Salei pack had taken over the former home range of the Naabi pack, which had been the main study pack before it died out in 1988. Lelo had the use of the only project vehicle, a small four-wheel-drive Suzuki that had been used by Fanshawe but was far from suitable for the Serengeti terrain.

During 1989, Lelo's main tasks were to radio-collar any new pack or group of wild dogs that could be located and to facilitate tourist access to the Salei pack's den, located just north of Naabi Hill. At the same time, he was to prevent any tourist vehicle from approaching too closely. Borner had a policy of keeping park administrators informed of the location of wild dog dens, which were major tourist attractions. In mid-1991, when only one pack—a lone pair with four pups—was known to be denning within the park, Roger suggested to Borner that frequent visits by tourist buses could have adverse effects on the dogs. Borner insisted that the location of the den should not be concealed. This small pack was the last seen alive in the park, as documented by tourist photographs, in June 1991.

Borner, a Swiss biologist who had done his doctoral work on the Sumatran rhinoceros, had no clearance from COSTECH for wild dog research in the Serengeti. He was the FZS's representative in the park, and the only representative of a foreign organization to sit on the board of directors of TANAPA. He was, perhaps most notably, also active as a pilot in the Serengeti. He had made the tracking flights that monitored the wild dogs' movements. Roger saw the need to do more on the ground; he was particularly interested in behavioral studies.

After Steven Lelo had shown Roger where he had found wild dog packs and dens, Roger and Jan began staying out for a week or more at a time after a pack had been located through radio-tracking. They studied the packs intensively, particularly the Salei pack. Only in this way could all the individuals be identified and photographed, thus enabling accurate counts and descriptions of changes. Roger made two sets of the I.D. photos so that he and Steven could separately identify the dogs when they located them on the plains.

"From February of 1990, we really got to know those dogs," Roger recounts. "We realized that there was a lot to be done with den study but an awful lot to be done with movement, so we followed them night after night, all night."

Wild dogs, nomadic by nature, travel fifteen to twenty kilometers (nine to twelve miles) each night when they are no longer tied to the den by very young pups. A "grand route march," as Roger describes it, can cover sixty kilometers (forty miles) in a few days, often during the period of the full moon, with female yearlings usually leading and the two- or three-month-old pups trailing behind. Shortly after a long-distance movement, the female yearlings begin to disperse; male yearlings usually stay a few more weeks or months in their natal pack before they, too, emigrate, some arriving eventually in Kenya. The beta pair may also break off at this time. This makes it difficult to consider a pack as an enduring group; it is flexible and dynamic, and its members may emigrate to form new packs. Immigrants, particularly young females, may also join existing packs. This is why it is so important to be able to identify the dogs as individuals and not just by the number that den, breed, or travel together as a pack.

The wild dogs' mobility may also facilitate transmission of disease from one dog or pack to another, although Roger points out that the majority of packs that may have been wiped out by disease included no recent immigrants and that none of the young radio-collared emigrants from the study packs died during their extensive travels or within a short time after they had formed a new pack.

A wild dog is not a domestic dog gone wild but a distinct and distantly related species. The Masai have long lived peacefully among them and know they are not dangerous to humans; more recently, ranchers have shot the wild dogs, fearing that they might prey on livestock, but no basis for that fear has been documented. An adult wild dog is about the size of a German shepherd but with longer legs and generally a sparse coat containing long guard hairs and few underhairs. In some individuals, this results in a mangy look, but others appear to have beautiful full coats. A white-tipped tail is a characteristic but not invariant marking; large bat-ears may be their most outstanding feature and, like the tails, may be used to signal. Pups, which are born naked, soon develop black hair and, as they mature, a more variegated coat with colored patches. As implied by its Latin name, *Lycaon pictus,* the adult animal resembles a painted wolf; each dog displays unique color patterns with mottled black, brown, yellow, and white patches. The dogs become drabber with age, and their ears grow ragged from packmates' playful bites.

Their excellent eyesight, notable endurance, and impressive speed make them outstanding hunters and account for another common name: African, or Cape, hunting dogs. A pack is led socially, but not always during hunting, by the dominant or alpha pair, often supported by a beta pair and sometimes other adults, usually brothers of the alpha male or sisters of the alpha female. The yearlings, twelve to twenty-four months old, are more active and are usually the fastest but not always the most effective hunters in the group. At night and during wet weather the dogs spend much of the time huddled together. In the heat of the day they seek the shade of isolated trees, lie in temporary pools of water, or retreat into underground dens.

Although there is normally only one denning period in the year, during the Serengeti wet season, both alpha females and subordinate females, usually sisters of the alpha female, can produce litters. Thus a pack can produce several litters in one denning season, and with up to fifteen pups in a litter, a potential exists for huge packs to form rapidly. Pup mortality is often high, however, and single litters larger than twelve and combined litters larger than twenty are rare. The alpha female whelps first; litters born one or two weeks later to her subordinates may not survive.

The dogs are extremely social, living in close-knit communities with frequent interactions between individuals and among pack members. Females, usually subordinates, will suckle pups that are not their own.

Wild dogs have a distinctive "hoo-call," a plaintive call to locate other members of their pack when they are separated. They bite and lick one another often, especially in greeting; the pups play much like domesticated puppies. Although wild dogs are individually capable of killing prey as large as male Thomson's gazelles, they also cooperatively hunt larger prey, such as adult wildebeest and zebra. Lactating females, particularly the alpha, are fed in the den by the rest of the pack. The dogs are almost exclusively carnivores.

Roger describes a "windup" for a hunt: in the early morning or about an hour before sundown, a dog rushes into the pile of sleeping packmates, jolting them awake and stirring up excitement. They begin to leap and stretch, urinating, defecating, loosening up for action. The dog that began the windup, often a hungry subordinate male, then shoots off in search of prey, followed by the others. Sometimes they all settle down again, but within a few minutes, another windup will begin. When they do home in, they approach the herd of prey cautiously, heads down, ears laid back, and single out an individual. Then, making curious birdlike twittering sounds, they begin to run it down. Their attack on a wildebeest calf or young zebra is quick and gruesome; the dogs pull the prey apart and begin to eat it sometimes even before the animal is down.

If the prey is large, all the dogs begin to feed once it is downed. Only very young pups, which stay behind in the den, are not at the kill.

After the first minutes of feeding, a hierarchy begins to assert itself, with the older adults deferring to the yearlings or older pups, which then have the carcass to themselves for the next hour or so. The subordinate adults wait until the yearlings have fed before they return to the carcass. The alpha female, after quickly eating, returns to the den to wait for the others to bring more food for her pups from the kill. Sometimes the alpha female or the beta female will stay in the den with the pups while the kill is being made.

Using what Roger refers to as a conveyor-belt system, the yearlings return to the den and are greeted by the pups, which solicit food by making twittering sounds and jumping up at the yearlings' mouths. As if on cue, the yearlings regurgitate their food for the pups. The amazing thing, Roger says, is that even after three or four hours, some of the yearlings are still able to regurgitate fresh meat. The pups eventually eat so much that they can barely move.

After the yearlings and the subordinates have eaten their fill, the jackals, hyenas, vultures, and eagles move in on the remains. One wildebeest may feed a pack of twenty-five wild dogs, with a good deal left over. Eventually the pack may move for a while to a nearby water hole, leaving the pups either by themselves in the den or with a couple of yearlings as baby-sitters.

Working from data of varying quality and completeness collected beginning in 1965, Roger began a detailed reconstruction of the life histories of all the Serengeti study packs. Some packs survived for many years before being reduced to a nonviable pair of old dogs, but before 1986 none died out suddenly and unexpectedly. After 1986 entire study packs suddenly disappeared.

In June 1986, three months after a subordinate male in the Pedallers pack had been radio-collared, members of the pack were reported to be dying. It was not long before the last of the Pedallers—the radio-collared male—disappeared. His empty collar was found on the plains and was later fitted to another dog.

The Naabi pack was then the only one left on the plains. This pack had been monitored from the air and frequently spotted by tourists because it stayed near the main road and park gate. Roger learned from the aerial reports that its numbers had been declining. By September 1988, four months after the sixth dog in the pack had been radio-collared, the pack had disappeared. One empty radio collar was found.

"I thought this was very strange that you would have this repetition of the Pedallers pack with a radio-collared dog and once you find the radio-collared dog's dead, the whole pack's dead. And the same thing happened with the Naabi pack. They monitored it from the air—gradual decline, found the empty collar, the pack's gone."

In 1989, Roger saw one pack, the Ndutus, but only briefly. "I only saw them with Vicky and Mark in August when they bred," Roger remembers. "I saw them briefly for an afternoon, and when I came back to start my research, they were dead." This pack had been formed in February of 1989 and included a radio-collared female that had emigrated from the Naabi pack. At the time the pack had formed, a male was also radio-collared. He left the pack and was dead by August. Another male was collared in July of 1989 and died, along with the rest of the pack, by December the same year. No carcasses were found, only two empty radio collars.

On February 23, 1990, Roger was with a group camped at Naabi trying to locate dogs for a BBC crew to film when they discovered a lone male which had joined two females that came from the Aitong pack in the Mara—120 kilometers (seventy-five miles) distant. The two females had been photographed in the same area a year earlier, and Roger had been able to identify them from a photograph of some of the 1986 Aitong pack pups on the front cover of Jonathan Scott's book *Painted Wolves,* the only source for photographs of the Aitong pack available to the Serengeti team. They christened these three the Lemuta pack; the male was collared by Laurenson and Lelo. That was the first and last time this pack was seen.

None of the dogs in these four packs—Pedallers, Naabi, Ndutu, Lemuta—had been vaccinated. All the dogs, however, had been handled when they were fitted with radio collars.

Dogs from the Naabi pack, including a radio-collared female and two males, had dispersed the year before the pack disappeared. They joined up with two unknown females and formed the Salei pack in

1989. This then was the principal pack Roger had to work with when he began his research.

One other known pack, the Ndohas, were in the park's western corridor, an area difficult to reach because the roads are sometimes impassable. Dogs there had usually been monitored by Borner and Laurenson from the air, but Scott had photographed and observed them on the ground in 1986–88. Unfortunately, no color photographs of this or any other pack were available, and the black and white photographs were of limited value. Later in his study Roger received a family tree of the study packs from Scott, which was very useful. It confirmed many events that Roger had painstakingly deduced from scraps of data.

For the five radio-collared packs monitored from 1985 to 1989, no I.D. cards existed for three and only incomplete sets for the other two. Of all individuals catalogued by FZS since 1985, only nine that appeared on I.D. cards were still alive in January 1990. Thus Roger found it hard to believe when he read in IUCN's 1990 pamphlet *Action Plan for the Conservation of Canids* that the work by Laurenson and Borner involved "keeping a photographic file of each individual in the population . . . and monthly radio-tracking flights to locate packs on a regular basis."

Roger's requests to get Borner to take him on tracking flights were frequently denied, although records showed that many flights had taken place before he began working with Lelo. Still, whenever VIPs visited the park, flights were quickly arranged to "see the dogs." After Roger reminded Borner that their agreement stipulated that Roger would make at least one flight each month, Borner told him, "You've really got to twist my arm." As flights to locate wild dogs were so infrequent, he resolved to make the best use he could of any flights that Borner undertook, which were mainly to locate radio-collared lions and cheetahs. If wild dog collar signals were picked up from the air, Roger immediately set off on the ground to locate the pack and stayed with it until he lost them, usually on moonless nights. At other times, he drove all over the park, learning the terrain and watching for any signs of dogs.

"I was worried, obviously, with the loss of the third pack but then I thought, well, I had no experience of this before, I thought this is perhaps what happens with wild dog packs. Perhaps they are ephemeral groups that come and go." He remained unsettled, however, about the methods used to evaluate the losses. "I thought it was rather strange that they hadn't gone down to collect samples. There wasn't a single tissue sample other than those taken when they radio-collared and took blood samples. And nothing was published about that."

During the denning season, observations posed no problem because the dogs stayed near their dens, but when they started roaming, it was difficult to find them without radio tracking. Even this expensive technology brought little of value if no tissue samples were taken once the packs were located. Because few observations were carried out on the ground of packs located from the air, dying and dead animals were not examined, nor were tissue samples taken. For just that reason, little had been learned about the reason for the disappearance of the three packs that had died out between 1985 and 1989. Yet, Roger reminded us, "tissue collection is one of the chief reasons given by scientists for radio-collaring wild dog packs."

Borner had applied to TANAPA for permission to vaccinate wild dogs in the park against rabies after Roger had told him of Pieter Kat's strong recommendation; TANAPA granted permission, even though at the time there was no confirmation that the wild dogs of the Serengeti had rabies. Karen Laurenson was also a vet and Roger deferred to her support of the vaccinations.

"Not being a vet, I did not know the questions I should, in retrospect, have asked concerning the advisability of vaccinating free-living wild dogs. Why should I, in 1990, have questioned what I took to be a well-thought-out strategy for attempting to conserve the dogs?" Roger asks. "I didn't know the background of Kat. He hadn't told me what had happened in the Mara or of the suspected rabies epidemic. I assumed that this was well researched, that you needed to vaccinate wild dogs against rabies, as there was an imminent rabies threat in the Serengeti also. We know that the Pedallers had died in 1986 of suspected rabies, as is reported in the literature."

Roger was also in contact with Hugo van Lawick, who told him of another pack of wild dogs, the Mountain pack, which he had been filming in the remote Gol Mountains of the Ngorongoro Conservation Area. Van Lawick hadn't said much about this group, Roger notes, because he had wanted to make a film of wild dogs without radio collars. When van Lawick's filming was nearly finished, in June 1990, Steven Lelo radio-collared two of the dogs in the pack, an old male and a subordinate female. Roger was with van Lawick when the dogs left their den—normal behavior for wild dogs—and began their nomadic trekking in a home range of probably more than a thousand square kilometers (about four hundred square miles).

Eager not to lose this newly discovered pack, Roger asked Borner for a flight to locate them a month later. After locating the dogs from their radio signals, Roger could see fewer dogs from the air than he had observed with van Lawick—they were small and therefore young dogs. From the air, as they banked to circle a kopje, he could discern no adults. The dogs were playing with something that looked like "a big piece of red plastic."

Back on the ground, Roger, still puzzled by what he'd seen from the air, decided to investigate further. He made the five-hour drive from Seronera to the place they'd seen the Mountain pack. "We found the pack, what was left of it, and it was the pups who were eating the remains of one of the dogs, an adult female," he reports. The adult, a subordinate female, named Snow White by van Lawick, could easily be identified, for her head was intact. "Her eyes were very bright, she'd obviously only recently died." Besides the pups, there was one adult dog, the old male which had been radio-collared by Lelo. He was lying in the full sun, peculiar behavior for a wild dog. There was no sign of the radio-collared female or any other adults.

Roger quickly made a video of what he saw and drove back to Seronera where he showed it to Karen Laurenson. She immediately suspected rabies had struck the group. Laurenson, Borner, and a newly arrived researcher and vet, Sarah Gascoyne (who later changed her name to Sarah Cleaveland), flew out the next day to the area where Roger had observed the Mountain pack. Roger, driving his Land Rover, met them as they landed; together they found Snow White's carcass, but no sign of the pups or the old male. Using Roger's screwdriver, a hacksaw, and a hammer, they opened the skull and took a straw sample from Snow White's brain. Before leaving the site, they found the collar that had belonged to the female collared by Lelo in June.

It had been about twenty-four hours since Roger had first seen the pups playing with the "red plastic."

Borner relocated the old male from the air, five or six kilometers (three or four miles) away, heading in the direction of the natal den. While Borner flew the vets back to base with the Snow White sample, Roger picked up the radio signal on the ground and followed the male throughout the day, making videos and recording clinical data. He recalls that the male would walk, stop, sway a bit, then sometimes lie down on his side in the full sun. The dog never panted.

Having followed him all day to the natal den, Roger stayed in the Land Rover when the male went underground into the den as darkness fell. Expecting that the dog would stay there overnight, Roger set his alarm, lay his head on the steering wheel, and fell asleep. "And I thought, he'd be there in the morning. If I'm up at four o'clock, no problem. If he's still alive, he's going to come out of the den; if he's not, he'll still be down there."

When he awoke the next morning, there was no sign of the male and no radio signal. Waiting until daybreak, Roger explored the den and found nothing. Further exploration of the area, including a nearby satellite den, also turned up nothing. Discouraged, he returned to Seronera and contacted Borner, who made a flight the next day in search of the old radio-collared male. Roger drove out again—the third time within four days—to be on hand if the dog was sighted from the air. Nothing was seen or heard, either on the ground or from the air.

The last of the Mountain pack had disappeared. It was August 1990, two months after the female in the group had been radio-collared. And it was the fifth study pack to disappear since 1985.

Vaccinations against rabies started in the Serengeti on September 1, 1990. Coincidentally, the same week confirmation came through that at least one dog in the Mountain pack had died of rabies. The result was based on the single straw sample that Laurenson, Gascoyne, and Roger had taken from Snow White.

The Salei pack was the first of the two remaining packs to be vaccinated. Roger and Steve Lelo were stationed in the Land Rover on the ground; Borner, Laurenson, and Gascoyne flew in with the vaccines. "The object was to vaccinate all of them," says Roger, including the adults, the yearlings, and the pups. "The idea was also to take the collar from the male in the pack—the collar had been on for a long time, since 1987, and its batteries were dead. But we were also going to radio-collar another younger male in its place at the same time the vaccinations were taking place." They took blood samples so there would be both pre- and post-vaccination blood samples on file. A young female in the pack had been radio-collared the previous January.

Roger describes how the dogs were immobilized and vaccinated. The Land Rover would move up to about twenty meters (sixty-five feet) from the dog that was to be darted. Laurenson fitted the dart filled with the tranquilizers ketamine and xylazine into the gun; Lelo decided on the correct setting for the air-pressure gun, aimed to hit the dog in the shoulder, and shot. The dart itself had to be pressurized so that the tranquilizer was delivered on impact. The dog normally went down within two or three minutes of impact. Lelo, Roger said, was an excellent shot and had previously immobilized dogs to radio-collar them without a vet present. Usually he needed only one shot. The dog that was down was then screened from the rest of the pack, usually by moving the Land Rover between it and the other dogs. "They wouldn't go away: they were curious. They would move a few meters away and attempt to see what was going on, and perhaps hoo-call, give the distress call."

When the animal was immobilized, the vet would put a cloth over the dog's head, the dog would be weighed and measured, checked for ectoparasites and skin problems, and a blood sample would be taken from its leg. If the animal was to be radio-collared, a collar with an external aerial would be attached and the frequency of its signal would be recorded. Frequently old collars that had been used on other dogs, some of which had died, or on cheetahs, were reused. Then the dog was given the rabies vaccination by hand, followed by a dose of a broad-spectrum antibiotic and an antidote to the anesthetic. "And then you back off and wait for the dog to recover," Roger concluded.

This is the procedure they used on the first male. The second dog, the male already radio-collared, was also immobilized and hand-vaccinated, and his collar was removed. The female that had been radio-collared in February was not immobilized; it was assumed that the sample taken in February would stand as a pre-vaccination sample, even though it is generally recommended that a pre-vaccination sample be taken on the day of vaccination before the vaccine is injected.

The rest of the dogs were dart-vaccinated, with the vaccine replacing the tranquilizer in the dart. Roger, the only one of the team who could identify the individuals, directed Lelo to each dog and recorded each vaccination on the animal's chart. The dogs became nervous during the long process and began to move.

"They became somewhat agitated by the process of having two dogs knocked out, the presence of two Land Rovers, and an aircraft not far away," Roger explains. "So they're moving around, while they should be lying down because it's getting warm."

The work continued slowly, Roger directing the driver toward the next dog, Lelo taking aim and shooting. It took about eight hours to vaccinate fourteen adults and the five alpha pups and was near the end of the day when they finished. They didn't vaccinate the beta pups, two months younger than the alpha pups, because Roger thought they were too small. The animals that were dart-vaccinated did not receive the tranquilizer or the antibiotic.

How traumatic was all this to the dogs?

"Well, the visual evidence of trauma, if you can call it that, is fairly slight," Roger replies. "You see the dog jump, yelp. The dart should fall out, but it usually didn't, it would remain, so the dog would be looking around to see what was going on behind it and trying to get at it. The other dogs would be fascinated with this red tassel on the end of an object sticking from one of their compatriots.

"The others in the pack would tend to go for it immediately when it hit the other dog. Sometimes this might have meant that the vaccine hadn't got through, but also they obviously chewed the plastic darts. We had a limited number of plastic darts and this eventually began to cause problems because we were running out of darts."

Borner, Laurenson, and Gascoyne flew back, and Lelo also returned to Seronera. Roger monitored the dogs the next day and into the night, when the dogs moved on. He noted no further reaction to the handling.

A similar procedure was used with the other study pack, the Ndoha pack in the western corridor, two weeks later. Because Roger had observed these dogs only once before, however, he first spent two days making identification drawings and taking descriptive notes. The pack consisted of twelve adults and four pups. Two dogs were immobilized and hand-vaccinated: a young male and a young female, which was also radio-collared with the only remaining collar, a second-hand cheetah collar. One of the other males in the pack had already been radio-collared. Again, the work took all day. The four pups were not vaccinated that day, because the team ran out of time; Roger and Lelo intended to vaccinate the pups the following day.

The airplane left, carrying Borner, Laurenson, and Gascoyne. Laurenson left the Serengeti for the United Kingdom the next day, and Gascoyne took over the vaccination program.

Roger and Lelo found the dogs in the morning, but the pack quickly disappeared, probably having crossed the river, and could not be followed. The four pups were never vaccinated.

Three weeks later Roger located the Salei pack and, together with Gascoyne, vaccinated the beta pups. Before the end of September 1990, the entire Salei pack and all but the four pups of the Ndoha pack had been vaccinated against rabies. In February, 1991, Borner, Gascoyne, and Lelo flew out to try to get a post-vaccination blood sample from the female in the Ndoha pack that they had radio-collared and to vaccinate the pups. They returned without having been able to obtain the sample or to make the vaccinations; they had, however, been able to radio-collar a young male. The collar they used had been taken from the subordinate female of the Mountain pack that had died of rabies in August 1990, for there was no other usable collar available. No photographs were taken of the young male. Roger, who was in England at the time, does not know which member of the Ndoha pack was radio-collared in early 1991. Gascoyne reported that there were new pups in the pack; Lelo reported that the alpha male had disappeared.

It was the last time the Ndoha pack was seen.

Roger, concerned about the Ndohas, later drove to the western corridor and discovered that the rangers there hadn't seen the pack since February. In May 1991, he arranged for a flight with Borner to locate the pack. He and Lelo waited two days at a campsite in the western corridor for the flight. Borner flew over Roger's camp on his way to Rubondo Island on the west side of Lake Victoria and radioed that he was too busy to look for wild dogs, he would do it later. Roger and Lelo then patrolled the area in the Land Rover but found nothing. About the same time, they tried to locate the Salei pack but were unsuccessful. Another attempt to rendezvous with Borner that month to locate the Ndoha pack did not pan out.

Compelled to scout on the ground, Roger enlisted other scientists in the park to help him, but the search was fruitless. The last time Roger saw any of his study packs was in June 1991, when he located the M & S, the two subordinates from the Salei that had chosen to den within a few hundred meters of a large hyena den. They had four healthy pups, the only wild dog pups in the Serengeti known to have left a den successfully in 1991.

Roger did see one lone female dog in late June, but she was not from a study pack. She may have been one of the wild dogs living on the margins of the park, which had not been handled or vaccinated.

While he was searching for the packs in June, Roger backed his Land Rover down to a dry riverbed to collect firewood. The gearshift first stuck in reverse and then, after being freed, jammed in second gear. With only second gear and some luck, he managed to drive the dilapidated vehicle back to mechanics in Seronera, a distance of more than eighty kilometers (fifty miles). Not all the needed repairs could be carried out there, and Roger decided that he would have to try to reach a Land Rover specialist in Arusha. On July 10, 1991, Roger and Jan limped into Arusha in the Land Rover for repairs that would take six weeks.

While waiting for his Land Rover to be repaired, Roger again looked over his data and reflected on them at length. "It was there in Arusha. . . that I went through the data and I looked at the amazing losses, starting in 1986, of five unvaccinated study packs, one with rabies suspected and another with rabies confirmed, all dying a few months after individuals in each pack had been radio-collared. And that was when I suggested to myself that there was something involved with the collaring process that I didn't understand."

He thought that certain stages in the life cycle might be particularly sensitive to stress caused by handling. In four of the five packs that had disappeared before the vaccinations in 1990, a post-disperser (an adult usually more than two years old that had left its natal pack and joined another pack) had been handled for radio-collaring. There appeared to be no problem with radio-collaring a dog younger than two years in its natal pack, before dispersal. It was difficult to sort out, however, because the whole pack was often vaccinated at the same time as a young dog was radio-collared. Both procedures could be triggers for stress.

He was still pondering the possible connections when Marion East and Heribert Hofer arrived in Arusha and informed him that four wild dog collars had been found in the park. Marion and Heribert normally stayed clear of Borner's team in the park but they had become friends with Roger and were concerned about the disappearance of the wild dogs.

"They're all dead," Roger thought aloud when he heard the news about the dogs' collars. He knew, he says, because every time collars had been found, the entire packs had died. "I just extrapolated from that."

The dogs remaining in the two vaccinated packs, the Salei and the Ndoha, had disappeared. In addition, two other packs formed entirely from former members of those packs could not be located. One consisted of single-sex yearling groups that had emigrated in late 1990 and had been roaming the Serengeti short-grass plains. Here in early 1991, the three former Salei pack females, one of them radio-collared, met up with three males from the Ndoha Pack, one of them, again, radio-collared, to form a new pack known as the Trail Blazers. The other new pack that had been lost comprised the breeding subordinate adult pair, the M & S, that had left the Salei pack.

One other study pack formed early in 1991 when five dispersing yearling males from the Salei pack, one of them radio-collared, joined a lone older, unvaccinated but radio-collared female to form the New Barafu pack. It was later enlarged when two unknown young females joined the pack. Thus three individuals in this new pack were unvaccinated. There was no news of this pack.

At a meeting with Gascoyne in the Mount Meru Hotel in Arusha, Roger learned some details: two collars had been found on the short-grass plains and two elsewhere. One was from the Ndoha pack, the others had not been identified. Gascoyne also had words of reassurance, though. She and Borner had recently flown over the New Barafu pack, which included two radio-collared dogs, in the Ngorongoro Conservation Area. "They're fine," she told Roger, who was pleased to hear it but remained worried because she had not checked the pack on the ground. "I thought this was amazing," Roger recalls, surprised they hadn't gone to look at the animals. In fact, Roger reflects that following this sighting from the air, the New Barafu pack was never seen again, nor were their radio collars ever found.

While in Arusha, Roger had a private conversation with George Sabuni, the acting director general of the Serengeti Wildlife Research Institute, the authority for which COSTECH had granted him research clearance to work. He told Sabuni he thought there was a handling problem with the dogs in the Serengeti. Sabuni was very concerned. "Why are you radio-collaring these dogs?" Sabuni asked him. Roger replied it was because it was the only way they could get a fix on the packs to locate them in the dry season and work out emigration patterns. "I was supportive of the concept of radio-collaring," Roger admits, but Sabuni's questions made him reconsider the wisdom of the whole concept.

He described to Sabuni how the Mara pack had died, and listed how the Serengeti packs went out one by one, after handling, in 1986, '88, '89, '90, and '91. It was known that rabies had caused the extinction of the partially vaccinated Mara pack in 1989 and had been responsible for the death of at least one individual in the unvaccinated Mountain pack in the Serengeti in 1990. But that did not explain why entire vaccinated packs had also died in 1991 in both the Mara and the Serengeti, with rabies confirmed in vaccinated dogs in a Mara pack. At the same time, unhandled packs and groups survived in the same ecosystem. Also puzzling was the fact that although rabies typically moves in waves of infection, no rabies outbreaks had recently been reported among other wild species, particularly jackals, which are susceptible to the disease.

"I did not think that all collaring was necessarily harmful," he explains. "It was the apparent association between a recent collaring event of a post-disperser and the disease-related demise within a few months of each study pack that hit me very strongly. I didn't understand how the events were related, maybe they were not, but once we put both handling and rabies into the same equation, I thought perhaps they were linked and provided a common reason why all the study packs in the ecosystem had died post-1985." Upset, but determined to understand just what had happened, he began formulating an hypothesis.

"The first step was to say, what did they die from? My answer was rabies.

"Second, if they're radio-collared, is this stressing them? And is there an association between stress and rabies? Is it possible that the dogs could be carriers of rabies, could have latent rabies, and that the stress of collaring a dog carrying a latent rabies infection—rather like herpes in humans—caused it to shed the virus in its saliva with the virus producing clinical disease in other pack members but not, at least at first, in the carrier individuals?"

Remembering that the old radio-collared male of the Mountain pack and the radio-collared dog in the Pedallers had been among the last to die, Roger hypothesized that they had probably been carriers, and when they had been radio-collared, the stress they had undergone might have stimulated the reactivation and release of the rabies virus.

Roger's hypothesis linked stress from procedures that involved handling—radio-collaring, vaccination, and blood sampling—with rabies. It was known that some wild dog packs had been exposed to rabies prior to vaccination and that they had significant titers of rabies-specific antibodies in their blood serum. Roger thought the stress of handling might be sufficient to reactivate production of the virus, which then spread through the pack and precipitated the wild dogs' demise.

"I was hunting around for something to fit all the data together," he explains. "And I'd be the last person to say this is right, but [those data] fit all the historical patterns that I could come up with, the history of each pack and their disappearances, and explained why, to me, the dogs had died. Now how you get from the stress of handling to rabies, I don't know for sure, but there seemed a correlation between the two."

It was a working hypothesis. Roger knew that at least some of the evidence at hand supported it and, in the normal scientific fashion, he wanted to prove or disprove it. He never thought that voicing such a hypothesis would bring him under personal and professional attack and result in having his permission to work in the Serengeti withdrawn by the chief ecologist of TANAPA.

That is what happened within the next three months.

Roger wrote a confidential letter to George Sabuni in mid-July describing his hypothesis. The same day he wrote to Borner and Lelo informing them of his meeting with Sabuni and recommended that "collaring should be halted until we have answers to the questions raised from the data." However, he waited to inform TANAPA until he could fully compile the data he needed. He then talked with local veterinarians in Arusha to solicit their input on the situation.

Roger wrote to Borner on August 14 to say he that had decided to leave the Wild Dog Project. He had been frustrated by not being able to get flights to track the wild dogs' radio signals and then concerned that collaring the dogs might not even be a good idea. He would continue to work on the dogs but would go his own way, he said. He had already written to Richard Faust of FZS in Frankfurt telling him of this decision.

Roger returned to the Serengeti at the end of August, to the surprise of many people in the park. The rumor had circulated that he had left in May when he had seen dying dogs but failed to report them to Borner. In fact, Roger had reported that two male dogs, one radio-collared and the other the alpha male, were missing from the Trail Blazers pack following the last tracking flight, but he could get no further tracking flight to relocate them. Later events showed that they were already dead.

Something else that was circulating upon his return was his "confidential" letter to Sabuni. Most prominent among people who had copies of it was Markus Borner. A letter from Borner and Lelo and critical comments on Roger's hypothesis were waiting for him. In the letter, they said they had "found" a copy of the letter to Sabuni. Roger discovered only later that his "confidential" letter had also been faxed to scientists around the world who worked on a number of different species in the Serengeti. Their comments on his hypothesis had been requested.

A few days later, Borner and Lelo came to see Roger, who was already packing up the radio telemetry equipment to return to FZS. Roger told Borner he didn't need the telemetry equipment, he would use binoculars. He also reiterated that he could no longer work with them on the project. To his surprise, Borner responded by suggesting Roger take over the whole Wild Dog Project. Roger could hardly believe Borner would give up control of the project, then he reflected on the implications. "I was going to be left as the fall guy," he remembers thinking. "It was going to be my project, my fault, nothing to do with Frankfurt—I could see this flashing up." Roger declined the offer. "I really told him what I thought about his whole setup and it left no doubt that there was no way I was going to work with him again."

Lelo mentioned during the discussion that some wild dogs still lived in the northeast. Roger had the same sensation of relief as when Sarah Gascoyne had told him in Arusha a few weeks earlier about seeing the New Barafu pack. "I thought, 'Thank God, there's a pack alive!'" Roger recalls. "All the questions I should have asked at that stage, I didn't ask."

They parted, Roger saying that he would carry on his work and that Borner and Lelo could continue separately. "So at that stage, we were talking together, albeit at a sort of distant level."

For the next three weeks, Roger drove across the Serengeti looking for wild dogs, particularly those most recently sighted by Gascoyne and Lelo. He never saw any and never encountered anyone who had, recently. On September 29, 1991, Roger and Jan left the Serengeti and returned to the England to be with Roger's father, who was dying.

In September 1991, Pieter Kat confirmed that no wild dogs had survived in the Mara, either.

In December, Roger, still in England, received a letter from B. C. Mwasaga, the chief ecologist of Tanzania National Parks, informing him that Borner no longer wished to support and work with him and that he, Mwasaga, was withdrawing Roger's permission to work in the Serengeti. Roger notes, though, that before he had left the Serengeti the Tanzanian government (COSTECH) had renewed his research permit for another two years.

In January 1992, a letter from Mwasaga to Borner was circulated among researchers in the Serengeti, which referred to a recent meeting Borner had attended at the Tanzania National Parks headquarters in Arusha. Regarding the "revival of the African wild dog monitoring project," it stated that the director general "formally approve[d] the project to continue as per its setup and establishment but without the involvement and participation of Burrows." In closing, Mwasaga apologized to Borner "for all the inconveniences and embarrassment caused to you as a result of Burrows's scientific misconduct. We however, rest assured of your renewed effort and concern."

In September of 1992, *Nature* published a letter from Roger in which he stated that all the study packs of wild dogs in the Serengeti and the Masai Mara had died or disappeared and put forward his hypothesis about how handling may have reactivated the rabies virus in the wild dogs. At that point he didn't feel that he had sufficient evidence to maintain that all the wild dogs in the two areas had died, although he feared that was the case. His next major paper, written with Marion East and Heribert Hofer and published in the *Proceedings of the Royal Society* in June 1994, documented the extinction of the wild dog study population in the Serengeti. Another paper by the three in the same journal the following year stated that they could find no convincing alternative to Roger's original hypothesis.

These publications elicited an exchange of papers in prominent scientific journals, many written by scientists had who worked in the Serengeti. Roger's hypothesis was repeatedly challenged. A string of charges and countercharges were made, data submitted and questioned, measurements, methods, and statistics presented and challenged, conclusions disputed, and replies and rebuttals exchanged. Finally, in December 1996, one of the journal editors had had enough. "This correspondence is now closed," he wrote tersely at the end of a reply published in *Trends in Ecology and Evolution*.

In January 1993, Tanzanian authorities placed a moratorium on the handling of wild dogs throughout the country. A ban had already been placed on radio-collaring cheetahs in the Serengeti; only lions and hyenas could now be collared. Kenya had already banned radio-collaring. Later in 1993, David Macdonald from the Canid Specialist Group of the IUCN's Species Survival Commission issued a special statement on the effects of handling on wild dogs, in which he referred specifically to Roger's *Nature* article and called for suspension of any further handling of the dogs by humans.

But while debate raged in scientific journals, the general public remained largely unaware of what had happened to the wild dog population in the Serengeti. The subject had barely been touched on in popular writing about this vast ecosystem, hailed as one of the earth's most spectacular wildlife habitats, which had been among the first areas to be designated by UNESCO as a World Heritage Site. It took a long time for either the Tanzanians or the tourists in the Serengeti to realize that the dogs were gone. People "were just not informed," Roger explains. To his knowledge, Borner never acknowledged the extinction of the wild dogs in the park.

Five years after he left the Serengeti, Roger's perception of what happened to the wild dogs has sharpened, and he is also more keenly aware of the importance of recognizing what has been lost.

"It is incredibly sad," he says, "to see a species like that, a coursing animal, a pack animal which is a very wild creature, which has survived tremendous persecution by humans for so many years, been shot on sight by hunters and considered to be vermin, survive through all this and then disappear in an apparent conservation program that is designed ostensibly to find out why there is a decline of dogs."

The program, he believes, came up with few data of real conservation value, and he now sees no justification for much of the "haphazard, frequent intrusions into the lives of packs of an endangered species carried out post-1985 in the Serengeti and the Mara." He rejects the notion that in order to preserve the species, all that Wild Dog Project workers needed to do was radio-collar a couple of dogs in each pack so that they could be located and occasionally to dart the dogs and take blood samples from them to check for exposure to disease. "If the samples were, as is now known, not analyzed until years after the packs were dead or experimentally vaccinated, and if so few tissue samples were collected despite the loss of thirteen radio-collared packs," he wonders, "what was the point of the exercise?" It appears, he says, to have been a "high-profile charade" with fatal consequences for all the study packs.

Although the Serengeti dogs had been monitored since 1985, only one tissue sample was ever collected—the one he helped take from Snow White in the Mountain pack. "We lost all the Serengeti study packs since 1985 and what have we got? One tissue sample."

Throughout geological time, he notes, the earth has experienced natural extinctions. But he points out that IUCN studies show that the extinction of the wild dogs in the Serengeti-Mara is not at all typical of what is happening to the wild dog populations in other parts of Africa. In much of southern Africa, the number of dogs is growing. He believes the population "has the potential to increase in some places if the conditions are right." It is thus even more striking that it was in a national park that the wild dogs completely disappeared.

Does he think that it would have been better not to do anything than try to do something and, despite good intentions, do it poorly?

"In this context, yes, I do," he replies. Before 1980, "a tremendous amount of work was done by what I would call the old-fashioned naturalist approach." In this context, Roger cites work by Hugo van Lawick, James Malcolm, George and Lory Frame, and others. He recalls that during the 1970s, after two experiments with radio-collaring, it was decided that the procedure was a waste of time; the researchers found that they could get along fine without radio collars. Roger also came to that conclusion, after not being able to get flights to pick up the signals. With "this playing at technology," he explains, "you have this aircraft which can pick up dogs from enormous distances—it was amazing. But on the ground this radio telemetry was useless." He pauses and adds: "It was being used as a spectacle."

It angers him to think that the endangered status of the dogs was exploited after they were singled out as a flagship species for conservation in the Serengeti, and it saddens him that the image of the park has been tarnished. "When you get to the Serengeti, you think, right, you're in the middle of this world-famous national park, the most famous ecosystem, the place you've always wanted to come since you were so high"—with his hand he indicates the height of a child. It was a shock to find that the people in charge, though they may have been well-meaning and interested in animals, had financial and career interests that far outweighed their scientific ones.

He felt sorry for the younger scientists, "amazing young people coming into that park, who, unless they integrated themselves with one organization and paid obeisance to that organization and the leader of that organization, were finished, and there was no way that they could get on at all." Talks with many of the younger researchers who have been in the Serengeti and elsewhere have been reassuring, though. "They have said, 'We're very glad you did what you did—we couldn't do it.'" Many, whose research position, doctoral work, and future in academics were at stake, didn't feel free to speak out.

As for himself, he says, the backstop was always: "I don't need to do this, and they can't do anything about my career." He pauses to make sure each word sinks in: "It doesn't matter because I doubt the integrity of some of the people that I've been dealing with."

As for the scientific debate surrounding his hypothesis that handling might endanger wild dogs, Roger thinks many people had just not thought much about the issue. "There are some very good reasons for doing it," he admits. The use of radio collars on cheetahs, lions, and hyenas was widespread in the Serengeti, as it is elsewhere. If radio collars were banned, research would become much more difficult. Some scientists' entire research projects hinge on some form of handling animals—immobilizing, collaring, or vaccinating them.

On the one hand, Roger believes, it was to be expected that his hypothesis about handling would provoke resistance. On the other hand, he points out, it was not a new idea. From an extensive search of the literature he has learned that the rabies latency and resurgence is an established concept. "This is what I found so amazing, that it was suddenly not acceptable to talk about latency."

Since 1991, Roger believes, his critics have carefully constructed a false explanation for why all the Serengeti wild dog study packs became extinct after 1985: "They were, they claim, a small, isolated population which had been in [an inevitable] decline since 1970 due to high disease mortality. This mortality was exacerbated by increasing competition from a rising spotted hyena and lion population. The final extinction of all the study packs throughout the ecosystem in 1991 was due to a virulent epidemic, 'a stochastic event,' and there was no reason to look any further as to the cause of the loss."

He finds this alternative explanation implausible and riddled with discrepancies. "In both sectors of the ecosystem, unhandled wild dogs survived. The question which must be asked but which has been avoided by those who falsely claim that all wild dogs disappeared is, how did the nonhandled study packs, groups, and individuals survive whatever killed all the study packs?"

He admits that the experience has made him cynical about the conservation movement in general. "I had felt, in my naïve way, that everyone in conservation was in it because they were really concerned with the animals and they really wanted to do the best they could for the animals." When he expressed his disillusionment to an editorial assistant at one of the more prestigious scientific journals, her response was, "Where have you been?"

"The question that I've asked myself many times," Roger muses, "is, whose animals are they? Whose dogs were they for me and others to research? Why was it that I and others suddenly arrived in the Serengeti and did what we did to those animals? . . . It appeared to me that it was a very haphazard event."

that he was seeing the dogs before he ever picked up a signal from their radio collars. During the first two months he had the wrong radio frequencies, but even with the correct frequencies, he was having difficulties. Marion took a radio collar, walked off down the road with it and before she was out of sight, Roger and Heribert couldn't hear the signal.

"He seemed to have a good heart; he seemed to be committed to what he was doing, was trying to do a good job but was being very much frustrated," Marion recalls about her impressions of Roger at that meeting. It was clear from what Roger told them that the Wild Dog Project had not been well organized and that the identification records for the dogs had been poorly maintained. "And as things went on and on throughout this project, we saw Roger getting deeper and deeper into the kinds off problems we have seen many other scientists have, of trying to have high standards, trying to do things properly, trying to account for money, trying to do everything as you should," she says.

"So, although the packs were radio-collared and continued to be blood-sampled and immobilized and more radio collars slapped on, he actually did his research on the ground as a behavioral study by simply following the dogs in exactly the same fashion as the original dog researchers in the 1970s," recalls Heribert.

Heribert explains that the signals from the radio collars would be picked up from the airplane much better than from the ground because the angle covered from the air is wider and the reception is thus better. While the signal from a good transmitter might reach three kilometers (two miles) on the ground, in the air it could be detected over thirty kilometers (twenty miles).

They watched closely what happened to the wild dogs, in the beginning because, like the hyenas they were studying, wild dogs were social carnivores that lived in the spectacular habitat of the Serengeti plains. Because Marion and Heribert were aware of the belief among some scientists that hyenas represented a problem for the dogs, they were interested in the interactions between the two species. In addition, the wild dogs were intrinsically intriguing from a scientific standpoint because they constitute one of the most social species to show reproductive suppression. Such suppression occurs, Heribert explains, in cases where although several adult males and females are potentially capable of breeding, only the alpha pair has offspring: the other animals in the group assist in rearing those offspring but usually conceive and bear none of their own.

The Max Planck Institute's first researcher in the Serengeti, Wolfdietrich Kühme, had written a paper in 1965 about wild dog social organization. Later Hugo van Lawick and Jane Goodall studied the Serengeti wild dogs; popular television programs and books were written about them, and they gained a high profile.

Marion and Heribert listened to Roger's technical problems and gave what advice they could. They began to accompany Roger and Jan to observe the dogs hunting and interacting with hyenas.

Marion and Heribert had been keenly aware that dogs were dying in study packs in the park; they knew that tour drivers and tourists had reported to Lelo and Borner that they had seen dead and ill-looking dogs. Entire packs disappeared. By June 1991, Marion says, the last dog they knew of had been reported to be "hoo-calling" on the plains to locate other pack members. "It seemed that he had lost the rest of his pack, so that should have been an indication that there was something going drastically wrong," she comments. The next month they drove to Arusha to give Roger the news that four radio collars had been found.

Heribert and Marion relate how they returned to their work in the Serengeti while Roger was waiting in Arusha for his Land Rover to be repaired. During that time they received an unexpected visit from Borner, who handed them a "report" that he said Roger had submitted. Borner asked them to comment on the report, because he had been asked by TANAPA to review Roger's findings. This "report" of Roger's was, in fact, the confidential letter to Sabuni that Roger had written a week earlier. It included his hypothesis about the effect of handling on wild dogs.

"When Markus asked us to comment on this, he pointed out that if this went through and was accepted by TANAPA, everybody's life in the Serengeti would become difficult because then radio-collaring of study animals would not be possible anymore, and so this was something we needed to think about," Heribert says.

They submitted a written opinion to Borner that Roger's report should be taken very seriously: that the data should be analyzed carefully and in detail.

facts were, as far as we can ascertain them, then they are written down and they are there for people to look at."

In March 1992, a Lycaon Population Viability Analysis Workshop was held in Arusha. It had been planned for some time, but because of the recent demise of the Serengeti wild dog study packs, it now seemed to Heribert and Marion especially relevant. It was sponsored by the Canid Specialist Group of the IUCN Species Survival Commission; Heribert and Marion were members of the Hyena Specialist Group of the same commission. Although not invited, they had asked to attend and were granted "observer" status.

In addition to research scientists from seven southern and East African and several non-African countries, national parks officials and several veterinarians took part. Heribert estimates that about fifty people attended the meeting.

Roger was in England and did not plan to attend the Arusha workshop, but he drove to the airport in Amsterdam to hand his data personally to James Malcolm, an American scientist on his way to the wild dog workshop with whom Roger had corresponded. Roger says it was the first that Malcolm had heard about the disappearance of the Serengeti study packs—and he promised to present the population viability analysis at the workshop. Neither those data nor Roger's findings were openly discussed at the conference.

Heribert and Marion wanted to be at the workshop not only because they knew Roger couldn't attend but also because they wanted to hear what other people who worked on wild dogs had to say about hyenas' impact on the dogs. "When we turned up at the wild dog workshop, I got the very distinct impression that our presence was not very welcomed by various members of the wild dog community," says Marion, "and I think that was solely because of this handling issue. They did not want to talk about it, they did not want to have to square up with what had happened here."

She and Heribert kept expecting someone to mention that wild dogs had gone extinct in the Serengeti. Finally, when Steven Lelo addressed the assembly, he admitted that no dogs remained in the Serengeti study packs. Questioned by a member of the Canid Specialist Group about whether he had seen any of the vaccinated packs during the previous three months, Lelo said no. Asked if he thought they were still alive, he also answered no.

Ten minutes after Lelo had finished his talk, Borner made a presentation. Borner, who had not been present for Lelo's talk, claimed he still had a study population of forty-two wild dogs. At this point Heribert questioned him about the dogs but says he didn't get any clear answers. Because Heribert was an "observer" and was not considered a wild dog researcher, he did not feel he could push the issue. No one else asked Borner to account for the discrepancy between his figures and Lelo's.

Gascoyne then presented a talk about the vaccination program in the Serengeti and reported that observation of the packs for adverse effects had been stopped four months after the packs had been vaccinated. She reported four deaths among the radio-collared animals.

"What I found most disturbing," Marion remarks, "was that a picture of everything being okay in the Serengeti was being projected at the wild dog workshop with the claim that a large number of wild dogs were still alive and being monitored. This myth should have been challenged by members of the Canid Specialist Group at the time, because there were many wildlife managers from Tanzania in that room and they were completely confused by the contradictory information they were hearing."

Following the Arusha conference, TANAPA officials asked the managers of all the research projects in the Serengeti to take senior park personnel to each of the radio-collared animals in their study packs, in order to ascertain how many radio-collared animals were in the park, where exactly they were, and how they were faring.

"I think the really important issue was that the Tanzanians wanted to be able to go and look at the animals themselves," Marion says. "We took the parks people out, the lion people took the parks people out—but you see, Markus couldn't take anybody out, because, of course, the dogs were dead. He didn't have a living wild dog then to take them to." Instead, she reports, Borner told the park authorities that it would be no problem for him to take the collars off the dogs and bring them in.

Marion and Heribert had left the workshop with some other insights. John Richardson, a veterinarian from Nairobi, showed a video of wild dogs in the Masai Mara that were confirmed afterwards as having had rabies. These dogs showed signs of what is known as "dumb" rabies, not the "furious" rabies that causes affected animals

Heribert and Marion believe that Tanzanians are worried about the international reputation of their country, as well as about the actions of scientists in the park, and that it is clear to the Tanzanians that something went wrong with the wild dogs, that many packs were lost in the ecosystem over a short period of time. "This has never happened before, anywhere in Africa—they know that," Marion states. "They think something is wrong, and they don't like people treating them like fools."

Heribert and Marion don't believe that scientists have the right to meddle with animals without first thinking the consequences through very carefully—including those of handling in any form. "If species are endangered, then to compound that effect by insensitive interventions by researchers, I think, is totally unacceptable," Marion asserts. The loss of the wild dogs concerns her all the more because she regards them as special animals. "Maybe I wouldn't feel quite so angry about it if it were the loss of a copepod or something."

If wild dogs were to become extinct throughout Africa, she continues, it would be "criminal," for they are irreplaceable; neither the role they play in their natural ecosystem nor that system itself is as yet fully understood.

Heribert speaks of "a sense of public responsibility"—not only for the dogs but for the Earth as a whole. "We have already influenced events on the planet so much that, for better or worse, any loss of a species is a testament to the lack of competence or the lack of will or the lack of whatever that would be required to arrange human affairs in such a way that everybody could coexist," is his comment. The usual human motivation for introducing changes in the natural order is personal greed, in his opinion, or occasionally the hope of improving the world. Conservation arguments made on the basis of economic sustainability, medical value, and so forth "all stop short of considering our total responsibility in this place."

Asked whether he believes that every species has a right to exist, Heribert suggests that a better formulation would be that people don't have the right to destroy. "If you don't have the right to destroy, you should let other creatures live. If you accept the right to destroy, then it says something very sordid about the human state of affairs. Why do we say that humans have the right to destroy plants and animals but they don't have the right to destroy other humans or conduct their lives in such a way that it affects others in a negative manner?"

Heribert and Marion no longer radio-collar hyenas, even though in their own work they have never seen indications that radio-collaring or immobilization have any effect. The old, familiar problem persists, however: they can't get tracking flights.

POSTSCRIPT

In 1998, there were reports that wild dogs were beginning to return to the Serengeti. Packs were sighted near the eastern and inside the southern boundaries of the park.

July 1996 issue—are at loggerheads in this story, in which developers and academics played leading roles on opposite sides. And another striking clash of interests: While citizens of the capital of Texas held town meetings and lobbied to protect a frail, 7.5-centimeter (three-inch) long, pale pink salamander and its home, one of their U.S. senators instigated a moratorium that prevented anything from being added to the Endangered Species List (ESL) nationwide for more than a year.

The Barton Springs salamander has lived where it does today for thousands, perhaps millions, of years, certainly since before American Indians swam in the springs, before settlers and ranchers arrived, before the town was named in 1840 for Stephen Austin, father of the Republic of Texas—at which time this town where the Colorado River crosses the Balcones Escarpment had 856 residents.

Today the population of the greater Austin metropolitan area is more than a million, and it is one of the most rapidly expanding urban areas in the United States. Barton Springs is still there, two hundred yards (180 meters) long and twenty feet (6 meters) deep. Dammed since 1929 but never chlorinated, it is filled with naturally crystal-clear spring water, which remains a nearly constant 68°F (20°C) year-round. Hundreds of thousands of people swim in it every year, adults paying $2.50 per day and children twenty-five cents during the summer; winter days, and evenings at certain times of the year, are free, as is swimming below the dam. Surrounded by trees and lawn, it is a lovely spot, "our classless country club," one regular calls it. On a hot Saturday there may be three thousand people swimming in Barton Springs Pool. Not only do more people use the springs as their local swimming pool than ever before—they increasingly live in the recharge zone and the contributing zone of the Edwards Aquifer, which feeds into Barton Springs.

Like many other native species, the salamander drew little notice until relatively recently. It first appears in scientists' field notes in 1946. In 1982, the U.S. Fish and Wildlife Service proposed the Barton Springs salamander as a candidate for federal protection under the Endangered Species Act (ESA), but no action was taken to put it on the Endangered Species List. Ten years later, in January 1992, two University of Texas scientists, Mark Kirkpatrick and Barbara Mahler, formally petitioned the USFWS to add the salamander to the ESL. Still no action from the government, but at public hearings in Austin, local residents stayed up all night to testify in favor of the salamander and the springs. In 1993, a paper by three scientists in Austin published in *Herpetologica* formally described the Barton Springs salamander as a new species, distinct from others in the region, and gave it a scientific name that signaled its precarious existence, *Eurycea sosorum*.

"The species is named in honor of the citizens of Austin, Texas, whose efforts to protect the quality of Barton Springs resulted in the passage of a citizens' aquifer-protection initiative in 1992," wrote the scientists. "This initiative is known locally as the SOS (Save Our Springs) Ordinance, and its supporters are known as SOSers. The specific name *sosorum* is the plural mixed-gender genitive form of the acronym SOS." They pointed out that Barton Springs is the only known habitat for the newly described species, which "appears to have one of the smallest ranges of any vertebrate in North America, occurs in an area of extreme environmental sensitivity, and is highly vulnerable to extinction." They also said: "We see no conflict between human use of the pool for swimming and the continued existence of this species."

The authors were David Hillis, professor in the Department of Zoology at the University of Texas in Austin (U.T.), his graduate student Paul Chippindale, and Andrew Price of the Texas Parks and Wildlife Department.

To David Hillis, who has become an authority on the *Eurycea* salamanders of Texas since joining the U.T. faculty in 1987, it still seems puzzling that no one had described the Barton Springs salamander earlier. His undergraduate adviser at Baylor University, Bryce C. Brown, who had collected it in 1946, had sparked David's interest in the salamanders of the area, but David says a formal description is just one of the things Brown never got to before he retired.

Before a formal description can be written, much field- and labwork must be done; meticulous comparisons and measurements must be made to establish beyond a doubt that the organism is a distinct species. In the case of the Barton Springs salamander, the work involved electrophoresis, DNA analysis, computer calculations, examination of the existing literature, and a survey of its habitat, part of which David and Paul Chippindale did with scuba equipment.

dicates the health of the springs system in general. The disappearance of the salamanders would be a warning sign for decline in the quality of the springs. I think that is an especially effective argument in this case because there is so much community interest in the springs. Even people who don't care strongly about the salamanders care very strongly about the beauty of the springs and the quality of the water in the springs."

In hearings concerning Barton Springs or the listing of the salamander as an endangered species, David has pointed out that this is one of the very few species whose geographic range is restricted to the city of Austin. "It is certainly the only vertebrate for which that is true. And I try to make the point that you can't have a species that's any more Austin than this one is, to appeal to the local pride. . . . Austin is a very active community in terms of conservation and interest in biodiversity and preserving some of the local areas."

Asked to estimate the number of Barton Spring salamanders, David hesitates, reflecting (this is July of 1996) that it has decreased "dramatically" during the past couple years, and guesses it to be "probably in the hundreds." When he began surveys in 1991, a two-hour scuba survey would reveal 150 salamanders. But since then, the numbers have declined: about 8 or 10 are spotted over the same two-hour period. He and his colleagues no longer monitor genetic information on the salamander, he adds, because they are concerned that taking any samples might endanger the dwindling population. "They are frail, skinny little things, so I think even taking a tail clip from them in order to get enough tissue would probably be deeply stressful to the organism."

Drinking water for the city of Austin is taken from the Colorado River, but during some periods of the year the city's outtake comes from the point where the Barton Springs flow enters the Colorado. During those periods, Barton Springs contributes substantially to Austin's drinking water. South of the city, most of the homes have wells that tap into the Barton Springs section of the Edwards Aquifer. An estimated thirty thousand to forty thousand people depend on those wells for their water, and the number is growing because the 350-square-mile (900-square-kilometer) watershed that feeds the springs is rapidly succumbing to urbanization.

In David's opinion, the critical problem caused by development, one that threatens both salamanders and homeowners, is silt. Chemical pollutants are also a water quality concern, but "right now that's not nearly as much of a problem for the salamander as the siltation." The silt has already clogged up pumps for wells in the area and rendered them unusable.

Cobble that had offered a perfect habitat for salamanders in 1991 and 1992 had silted up by 1995. "We never saw any salamanders in those conditions, and so the only places we could find them were actually right in the springheads where the flow of water was sufficient enough to keep the silt from settling," David says.

What would it mean to him, if the Barton Springs salamander became extinct?

"Well, it would be a terrible tragedy. It would certainly decrease my pleasure in enjoying the springs. I go to the springs mostly to be able to swim around and look at wildlife. That includes not just the salamander but the other organisms that live in the springs. There are also a number of other endemic things which live in the same aquifer, so it's not just the salamander. The Barton Springs aquifer appears to have been fairly isolated from other aquifers for a long time. There is an endemic genus and species of snail that lives there—and since all these things are dependent on the same aquifer, the emphasis has been to take the organism that has the largest public appeal, namely the salamander, and try to protect that, because protecting the salamander will protect other endemics."

Realistically, he estimates the long-term probability of survival of the species to be very low. "Even with the best-case scenario, if we list it as endangered and we take strong actions to prevent degradation, I think continued development is still going to go on, and I think it is quite likely the salamander is not going to continue to be around a few decades from now, anyway, whether it's next year or twenty years from now."

Does he believe that the salamanders in Barton Springs have a right to exist?

"If you mean a right in some sort of natural scheme of things, sort of a natural right, then I don't see why humans have any more of a right to exist than salamanders do," he answers. "Certainly if you talk about humans' use of the springs, then in that case you could certainly argue that salamanders had use of the springs long before hu-

"In my mind, there's no doubt that this is an endangered species," David says. "It could blink out almost overnight if you had a fairly minor change to the ecosystem. If somebody comes along and bull-dozes a road right across a major recharge zone and you have a big flood, a big rainstorm, you'd have enough silt wash out to wipe out all the available habitat for the salamander overnight. If that's not im-minent danger of extinction, I don't know what is."

If the Barton Springs salamander is a fascinating endangered species in the eyes of scientists like David Hillis, it is an endearing symbol of the city of Austin in the eyes of local sosers, and a poten-tial giant-killer in the worst fears of developers and businessmen who want to see Austin grow. A little recent history will help the reader understand some of the intensity and complexity of the salamander's situation; anyone who wants to pursue the subject via Internet can do so (www.cs.utexas.edu/users/boyer/fp/).

For many who swim in Barton Springs or live in the area, the wake-up call came in 1990 when Freeport-McMoRan, led by Jim Bob Moffett, a former University of Texas football player turned miner and international developer, proposed a four-thousand-acre develop-ment in the Barton Creek watershed, tantamount to building an en-tire new city. When this proposal came before the Austin city coun-cil, it provoked a huge public outcry. Previous building in the area had been restricted primarily to single-family homes on large lots, but the Freeport-McMoRan proposal called for millions of square feet of commercial and office development, plus twenty-five hundred apart-ments and twenty-five hundred homes.

At what is now a legendary city council meeting, residents packed the hall on June 7, 1990, to listen to Freeport-McMoRan board chair-man Moffett and others defend their development plans. Protests, which were numerous, came from a variety of sectors. The meeting ran all night and at five or six in the morning the Austin city council voted 7–0: no.

One of those who attended the meeting was Bill Bunch, an attor-ney specializing in environmental cases. Born in San Antonio, south of Austin at the other end of the Edwards Aquifer, he grew up in the urban sprawl of Dallas–Fort Worth where, he remembers, one had to go miles to find clean water in which to swim outdoors. He credits his camp-outs with the Boy Scouts with stimulating his early interest in the environment. A competitive swimmer in school, he traveled to Austin for a swim meet when he was very young. On that trip he swam in Barton Springs for the first time.

After earning a degree in environmental biology at the University of Colorado in Boulder and attending law school at the University of California at Berkeley, where he took particular interest in environmental law, he returned to Texas and initially worked throughout the state on conservation matters. Drawn increasingly to the Austin area, he went out on his own as a private attorney focusing on Austin's land, water, and wildlife issues, doing a mix of volunteer and paid work, "but doing what I wanted to do." He acknowledges that he could be making a lot more money by doing other types of law, but "for people who are really committed to the environment and have a passion for it, I think it can be done, because there is an interest there, in the community, to protect the environment." Pressed to estimate just how much this passion costs him, he thinks for a moment and says, "I think I could have pretty readily been making seven to ten times what I'm making right now."

The city council meeting in June 1990, made it clear, he said, that the public did not want the proposed development, "that they believed this kind of development was completely inappropriate for Barton Creek."

"To the city staff's credit," Bill recounts, "in the course of that hearing, they admitted that the city's regulations in place at that time could not prevent pollution from that kind of development." The city council charged the staff with drafting an ordinance, which they did. It limited how much of the parcel to be developed could be paved over with impervious cover, such as parking lots, rooftops, and driveways. It also contained stream setback requirements and restrictions on building on steep slopes. This ordinance was adopted on an interim basis: by the summer of 1991, Bill and others could see that real estate interests were lobbying hard to water it down. There did not appear to be enough political support to implement a tough ordinance on a permanent basis. At that point, several environmental groups formed the Save Our Springs Coalition. "And we were very blunt," says Bill, who was on the leading edge of that move. "We said, 'Save Our Springs—if you don't, we will.' We were very up-front, we

month period was set within which a final rule listing the species was to be published. The deadline for that was February 1995, a date that also passed without any action by the USFWS. A month earlier, January 1995, the monthly census by city of Austin scientists had found only a single salamander in Barton Springs Pool, a record low. Time was running out.

After Mark and Barbara filed the petition to list the Barton Springs salamander, and after the citizens of Austin voted over-whelmingly in favor of the SOS Ordinance, another Texan made a move in Washington that was more than foot-dragging: it stopped things cold. In April 1995, U.S. Senator Kay Bailey Hutchison (R-Texas) added a rider to an unrelated bill before Congress that imposed a moratorium on adding any new species to the Endangered Species List. That moratorium, initially enacted for six months, lasted for over a year. Its effect was to halt listings of plants and animals throughout the United States.

There is no doubt in Bill's mind that developers were lobbying for the delays. "They were filing comments through the formal channels of participating in the listing process, but they were also exercising political influence outside that formal comment period," he says. Part of the difficulty, he explains, was that in 1989 a conservation plan had been initiated for all of Travis County, where Austin is located; it focused on the Golden-cheeked Warbler, the Black-capped Vireo and six other endangered terrestrial species. This Balcones Canyonlands Conservation Plan (BCCP) had been hailed as a way of allowing development while protecting the environment. It was credited to Interior Secretary Bruce Babbit as a path-breaking achievement and proof that the ESA was indeed a workable law. In a dedication ceremony in Austin in May 1996, a federal official signed a permit officially sanctioning the BCCP that would create a thirty-thousand-acre, $160-million preserve. The salamander was not mentioned.

If the Barton Springs salamander—which may be more at risk than any of the eight officially endangered terrestrial species—were added to the ESL, it could also be incorporated under the BCCP. Water quality regulations in Travis County would be affected. Developers with plans for building on the Edwards Aquifer or its watershed would face new obstacles, the Greater Austin Chamber of Commerce could find growth figures falling, and the interior secretary would face increased opposition to the ESA.

Bill sees the BCCP as a missed opportunity to formulate a comprehensive ecosystems plan and regrets that both the salamander and water quality protection were omitted from it. "And now they turn around and say, you can't surprise us with a new listing," he objects. "To me, there's no surprise. It's been a candidate species since '82." This "no surprise" policy—the concept that if a conservation plan is in place in a given area, newly listed species can't be added to its protective umbrella—is insidious, Bill thinks. "Nature, I guess, is not supposed to surprise us anymore. To me that's very disturbing."

Under the 1973 U.S. Endangered Species Act, if a species is listed, the federal government has a firm mandate to protect it. Bill points out that the act also includes the right of citizen enforcement: environmental groups and interested citizens can bring suit against those who are harming endangered species, to prevent that harm from occurring, and against the Interior Department for failing to list species or to protect listed species. The ESA is due to be reissued, and Bill is one of those who feel concern about what provisions the new bill may—or may not—include.

The citizen suit provision in federal environmental laws states that affected citizens may file notice of intent to sue and thus put an alleged violator and the federal agency charged with enforcing the law on notice that they believe a violation of the act is taking place. The violator is given sixty days to correct the problem or initiate action, and, if no action is taken, the citizen can bring enforcement action. "And that's what we've done with the failure to list the salamander," Bill explains. "Once a petition is filed, a series of deadlines kick in that are mandatory—there's no may or could or might or should. It's *shall,* within x months, take the next step." He adds that most states, including Texas, have no local laws to protect endangered species. Austin legislators, in fact, wrote such a law in the early 1990s, but it never was put into effect, because the state legislature passed a law prohibiting local governments from protecting endangered species.

In November 1995, Bill, acting on behalf of the SOS Alliance and Mark Kirkpatrick, filed suit against Interior Secretary Bruce Babbit to force a decision about the salamander's listing. Barbara explains that

addition, he asserts that taxpayers provided the water and sewer service to the Circle C project as well as another forty million dollars for highway construction to provide access to the development. Recently Bill has been adding up the public monies contributed toward the destruction of the aquifer, and thus toward endangering the salamander in its habitat, since the salamander was first identified as a candidate for protection. "We've come up with about nine hundred million dollars of taxpayer money going to destroy this incredible resource that the people of Austin have just screamed for the last twenty years they want protected." He suspects that another two or three times that amount may also have been spent that he has not yet identified, and he is enraged: "I mean, they're endangering this whole ecosystem, this water supply—and they're doing it with our money!"

Bill, like many of the other regular swimmers, has never seen a salamander in Barton Springs. He says he used to look for them but finally gave up: "I'm happy just to know they're there." The Barton Springs salamander, he feels, is part of home, part of Austin, and future generations would be done a great disservice if it disappeared. "I also feel that as the salamander goes, so the springs and the aquifer go, too—the salamander is important, but it is also a symbol and a warning signal that we're out of balance and we'd better pay attention."

Mark Kirkpatrick's office on the University of Texas campus, where he is professor of zoology, is not far from a large construction site where a new molecular science building was being built in the summer of 1996. He points out that the U.T. regents have approved naming the new building after Jim Bob Moffett and his wife, Louise, who donated two million dollars of the twenty-six million needed for its construction. Moffett, an alumnus and the head of Freeport-McMoRan, has maintained close ties with the U.T. administration. William Cunningham, the university's chancellor, sat on Freeport-McMoRan's board of directors until he resigned under pressure in December 1995, after the U.T. faculty senate passed a motion asking the regents to reconsider naming the building after the Moffetts. Freeport-McMoRan threatened to sue the three tenured U.T. professors who led the move, as well as Bill Bunch and three others who had been critical of Freeport-McMoRan. The Texas papers quickly picked up the story, pointed out conflicts of academic freedom and business interests, and questioned allegations about Freeport-McMoRan's reputation as an exploiter of mines in Indonesia and Papua New Guinea. Editorials appeared discussing the moral boundaries of the debate; the university administration and the faculty split over the naming of the molecular science building. University of Texas president Robert Berdahl, who had supported the regents' vote to name the building for the Moffetts, resigned at the beginning of 1997 to take up a seemingly less controversial position as chancellor of the University of California at Berkeley.

The controversy over the naming of the building, Mark says, really heated up when national media reported that human rights violations had been found to be associated with a mine in Irian Jaya owned by Freeport-McMoRan. A class-action suit was filed against Freeport-McMoRan by the Amungme people of Irian Jaya, protesting pollution from the mine and seeking billions of dollars in damages.

By comparison with its mines overseas, Freeport-McMoRan's proposed development in the Austin area is small change. From its Grasberg open-pit mine in the Sudirman Mountains of Indonesia, the largest single gold deposit in the world, Freeport-McMoRan takes $7.2 million worth of copper, gold, and silver daily.

Mark and Barbara were impressed that Austin's citizens had mobilized for the 1990 city council meeting and had formed several interest groups centered on protection of the springs. As Mark explains: "There is a diffuse and many-headed attempt in Austin to protect the springs, and the protection of the salamander is one of several fronts on which that battle is being fought. It is not the only one." He lists some of the groups active in the battle: the SOS Alliance, the Save Barton Creek Association, the Sierra Club, the Audubon Society, and Earth First. Additional steps have been taken to raise public awareness. For example, the Hill Country Foundation, a nonprofit conservation organization dedicated to protecting the water, wildlife, and scenic landscapes of the central Texas hill country, has published an "eco-location map" targeted for incoming residents, which shows the areas where people should avoid buying or renting homes. "By locating away from the most sensitive areas—over the Edwards Aquifer, in our water supply watersheds or in the habitat of endangered species—you can help protect our city's unique quality of life," the statement on the map reads.

More scientists than activists by inclination, Mark and Barbara decided to concentrate on gaining protection for the Barton Springs salamander. Mark gave priority to assuring good water quality in Barton Springs and requiring environmental impact studies on projects in the Barton Springs watershed. The highway projects already under way had begun without any evaluation of the impact they would have on Barton Creek, the aquifer, or the salamander. "That's just blatantly irresponsible," Mark charges. "There is absolutely no reason for that."

Listing the salamander under the ESA should halt development at least until impact studies show that it is not at risk. Mark says the Earth Day decision by Babbit to add the salamander to the list of endangered species came none too soon. "Recent studies of Barton Springs and the aquifer have found lead, arsenic, and gasoline components in the water and sediment," he explains. "At times, these toxins have been at levels that are dangerous to the salamander and to people. It's clear we have to work to reverse the damage that's already occurred and prevent further pollution."

Early on, Mark says, he sat down with David Hillis and told him that he was thinking about petitioning to have the salamander listed. David agreed that it was a good move, Mark says. "He didn't know, and I don't think I knew, quite what the whole story was going to be."

Pursuing the petition became time-consuming, certainly more than just something to think about during his regular swims in the springs. The preservation effort has set loose "a horrifying, fascinating, and exhilarating cacophony of science, politics, economics, and conservation." In the situation surrounding the springs and Austin's dilemma, he sees an encapsulation of much of the national, and even the global, biodiversity crisis.

In Mark's view the underlying question is, How much can local political will accomplish when a well-educated and committed community speaks out so clearly? "Should . . . we, with all our technology and resources, be able to do something, even on a scale as small as a 350-square-mile watershed sitting in the middle of Texas? If we can't do something that local in the middle of the world's most productive country, what are the chances for doing something global?"

Mark attributes a major influence on his thinking to his parents and their idea that "democracy is a participatory sport and you really need to be involved in your community, above and beyond going out to vote every once in a while." Barbara, he says, has reinforced that belief.

Barbara decided to go into hydrogeology only after moving from the San Francisco area to Austin. She also credits her parents with influencing her ecological thinking, primarily by instilling a strong appreciation of the outdoors in her through family hiking and fishing trips. When she was in high school in northern California in the 1970s, the politics of water and pollution was widely discussed. An interest in music led her to spend four years at Boston University, where she earned a bachelor's degree in music performance before returning to California to work as a freelance musician—playing the flute with small pick-up orchestras, teaching flute, and eventually becoming a member of the Monterey Symphony.

When she met Mark, her interest in the natural sciences began to grow, and with the move to Austin, her fascination with water issues revived. She completed a master's degree at U.T. in hydrogeology, working at "a beautiful spot" called Hamilton Pool, a spring-fed grotto in Western Travis County. The karst, watersheds, and aquifers such as the Edwards Aquifer, which feeds Barton Springs, captivated her, and today she speaks as passionately about karst aquifers as she does about music. Her doctoral thesis is on sediment transport in karst aquifers, a subject central to the problem the Barton Springs salamander faces.

"An aquifer is a body of rock which contains sufficient water so there is an economic value in pumping it out," she explains, warming to the subject of the Edwards Aquifer. In most aquifers the water moves through holes in the rock that were created when the rock was initially deposited. These are characterized as primary porosity. Secondary porosity develops after the rock is laid down, when something happens to create spaces in it through which the water can move. In karst—a limestone formation typified by caves, sinkholes, and highly porous rock—the slight acidity in rainwater and groundwater dissolves the rock itself. "Groundwater seeps into the little cracks in the rocks, so wherever there are cracks in the rocks, there is a zone of weakness where the water can get in, and it starts dissolving it away," she continues. "Instead of water moving very slowly through little holes that are a millimeter or less, it's moving through big tubes. And as you can imagine, the dynamics of the flow are very different."

These honeycombed rocks of the Edwards limestone become conduits for underground rivers, pulled by pressure or gravity, and the flow can be immense.

"The Edwards is interesting," Barbara continues enthusiastically, "because the porosity is aligned along a bunch of faults and fractures which formed during the Miocene, well after the aquifer. The aquifer was deposited during the Cretaceous period, say eighty million years ago, and then the faulting occurred much more recently. And it's faulted in a north-northeast-tending direction." If it weren't for the faults, she says, the water would flow east toward the Gulf of Mexico.

The contributing zone for the Edwards, she explains, while quickly drawing a sketch, is about 260 square miles (675 square kilometers). The area lies on top of the Glen Rose formation, which is much less permeable. The water flows east over the surface of the Glen Rose, but when it hits the fault where the Glen Rose is displaced by the Edwards, it is sucked underground, into the recharge zone, which is about 90 square miles (235 square kilometers). Instead of flowing toward the east now, the water suddenly enters pipelines that are basically heading north. "So the water does a ninety-degree turn, stops flowing to the east, starts flowing north towards Austin," Barbara says. "And it all comes out at Barton Springs."

The normal flow of water from Barton Springs is about fifty cubic feet per second, or twenty-seven million gallons a day. During a recent drought however, the flow was cut almost in half.

Forty thousand people living outside the Austin city limits, in northern Hays and southern Travis Counties, currently rely on the Barton Springs section of the Edwards Aquifer as their sole source of water. "And it would not surprise me if it that number doubled in the next ten years," Barbara comments. She is worried because recently she has seen huge numbers of wells being drilled in the recharge zone. "I've been talking to well-drillers, and they can't keep up with it."

The water coming out of the wells on the Edwards is so clean that it doesn't need to be chlorinated. "This is the great thing about karst water," Barbara says, her enthusiasm soaring. "Water out of karst aquifers tends to be—unless it has been polluted at the surface—naturally very, very clean water because it flows so fast that it doesn't sit inside the aquifer for very long and it doesn't pick up a whole lot of the constituents that are in the aquifer. There's not a lot of organic matter in karst aquifers because there's usually not a very thick soil cap." To illustrate how clean it is, she estimates that 80 to 90 percent of all bottled water comes from springs emerging from karst.

In the process of working on her doctorate, Barbara has looked carefully at the wells on the Edwards Aquifer and noticed something disturbing: they are filling up with silt. "Not all the siltation in the aquifer is actually coming from the surface," she says. "There are actually parts of the aquifer that are eroding internally and then being redeposited in other areas. Now, how much that is being affected by human activities is unclear, but it is very possible that some of the activities on the surface which are setting up vibrations are actually shaking loose some of the very loosely packed subsurface material." Karst, she emphasizes, is a fragile system. "There's not yet been a case where they have successfully cleaned up a major spill in a karst aquifer."

She recounts the changes in the area during recent years, including the building of the Barton Creek Mall on the recharge zone. At that point, she says, creeks nearby began to silt up, and water pollutants were traced to the parking lots. The old silt fences that had been built in the area no longer worked, so new silt traps were constructed. "They didn't work either," she shrugs. The runoff and siltation in the area have also been affected by the Texas Department of Transportation's construction of a new freeway interchange to handle the increase in commuter traffic. In addition, none of the highway bridges over Barton Creek have traps for hazardous material. According to Barbara, if a truck accident occurred on a bridge, oil might consequently spill into Barton Creek and come up in the springs.

Barbara notes that Austin area residents frequently tell her that they want to stop relying on groundwater and make a change to surface water. In California it was just the opposite. "You know, the grass is always greener on the other side," she says. "But the bottom line is that treating surface water is phenomenally expensive; delivering surface water is phenomenally expensive. We're talking ten times the cost of pumping water out of the ground."

In her estimation, the residents of the Edwards Aquifer area simply have no idea of the value of the service Mother Nature has provided for them. They also frequently underestimate the potential damage to the water system on which they depend. "My gut feeling is that you have this incredible, fragile resource here. Why mess it up?" she asks.

What would it mean if the Barton Spring salamander became extinct, because of siltation or a truck spill on a bridge or for some other reason?

"If we lose the salamander, to me it's an indication that probably the whole aquifer is going to go belly-up." That, Barbara adds, would affect more than a million people. "It affects everyone in the greater Austin area; it affects people who have never swum in the Barton Springs, who have never been to Barton Springs. The quality of life and, if you want to get crass about it, property values, are largely affected by Austin's physical environment, in which Barton Springs plays a key role. And, of course, it plays a much more important role for people who are pumping it out of the ground and drinking it every day."

The heat of the June day is beginning to ebb as Beverly Sheffield walks down to take a swim in Barton Springs. He stops often to greet old friends and new visitors at the edge of the pool. He is well known here, a virtual landmark. He was ten years old when he took his first swim in the springs in 1923. Tan and lean, he still swims several times a week, but not as often as he once did.

Beverly was Austin's first aquatics director in 1935. He trained lifeguards and supervised swimming pools, including Barton Springs. He became acting director of the Parks and Recreation Department before World War II, and director when he returned from active duty. When he was fifty, he realized that sitting behind a desk was taking a toll, so he joined the others who regularly swam in Barton Springs. That, he says, made a huge difference in the way he felt, and he has never stopped. He left the position of parks director in 1973 to organize Austin's bicentennial celebration, but he hasn't left Barton Springs.

He is proud that the springs was never chlorinated and proud that the salamanders are still there. He points to the Save Our Springs Ordinance and the 64-percent vote passing it as proof of how popular the pool is.

He has followed the lobbying and court actions and development plans that have affected Barton Springs. He calls Jim Bob Moffett's plans for the area "unreasonable" and has little sympathy for development on such a large scale. "He not only wants to make a profit, he wants to make a killing," Beverly says of Moffett. "There just needs to be more reason and sense to all this."

Beverly believes, however, that development is not the only factor endangering the springs. The flow of water itself is in danger because today about thirty-five thousand people are living off the aquifer, compared with fewer than two thousand just forty years ago, he remarks. "They're drawing their water from the aquifer to flush the toilets, to water the lawns, to take a shower, to drink, to cook with."

Beverly worries that if the water level in the Edwards Aquifer gets too low, salty water—which he calls "bad water"—from an adjacent aquifer will seep in and pollute it. Until now, he says, hydrostatic pressure has kept the bad water out.

"But we know that if they keep on with all this development, sooner or later, in my judgment, the springs are going to be polluted—really polluted."

He is somewhat pessimistic. "I want you-all to understand that this aquifer is a very fragile aquifer," he drawls to visitors at the edge of his pool. Water here flows too quickly through the honeycomb limestone, he explains, to rid itself of pollution of the type developers could cause.

Beverly has logged all his swims of over a quarter-mile, or one lap, since 1963. "After swimming the first hundred miles, I thought, well, I'll swim all the way to Waco," he recalls. "Then, here a few years ago, I said, well, I wonder how far it is all the way across the nation, and I picked from New York to Los Angeles and the map I used said 2,820 miles, and I swam that in 2,714 swim days." In the summer of 1996, he was aiming to swim as far as Hawaii.

He has a parting comment as he heads to the pool: "Barton Springs is more a state than a place. . . . You plunge in, you feel this beautiful water about you, and it's like heaven."

Raphus cucullatus. Drawing of a dodo by Lorene Simms.

IO

ISLANDS OF THE LIVING DEAD

There are ten species on Mauritius and Rodrigues which are down to one or two individuals, which are the real "living dead." But there are over two hundred more that have stopped reproducing in the wild, and many of these are on the edge of extinction.—WENDY STRAHM

 ONCE WE WERE fascinated by collective nouns for animals, sometimes colorful, sometimes whimsical, and often of ancient or obscure origin: "an exaltation of larks" captures the beauty and spectacle of a flock of singing birds climbing heavenward; *a pride of lions* is a term going back to the Middle Ages, and we longed to use it in the Serengeti; "a crash of rhinoceroses" describes, with power and humor, animals that are today among the most rare and endangered. When we met Wendy Strahm, she gave us a new and sobering collective noun: an island of the living dead.

This sad description refers to at least two islands, according to Wendy. They are Mauritius and its dependency Rodrigues, which, along with Réunion, make up the Mascarene group located about five hundred miles (nine hundred kilometers) east of Madagascar in the Indian Ocean. Wendy says that on these islands only one or two individuals each from at least ten species of plants are left in the wild, a status that dooms them to be "living dead," for none of these plants will leave progeny. It is a chilling phrase; together, these species make the most tragic collection imaginable. Even more disturbing is how many others are about to be added to the collection.

Speaking at her home on the edge of a vineyard, not far from where she works at the World Conservation Union (IUCN) headquarters in Gland, Switzerland, Wendy describes her first trip to Mauritius and the eleven years she spent there, from 1981 to 1993. After graduating in biology from Oberlin College in Ohio, she was awarded a Watson Fellowship for "focused wandering." An interest in conservation and wildlife and a longing to visit Madagascar led her to propose a project in nature conservation on that island. Her proposal was accepted, and she flew to France to brush up on the language first. Once there, however, she learned that it was impossible to get a visa to stay in Madagascar for more than one month, or in exceptional cases three, which wouldn't give her enough time to do her project. While she was considering alternatives, she received a phone call from Lee Durrell of the Jersey Wildlife Preservation Trust, who had heard about the problem. He asked if Wendy would be interested in working on a project on Mauritius. The proposal did not appeal to her initially because she thought of Mauritius as about 90 percent sugarcane. But faced with having to give up the Watson Fellowship, she called the people who had awarded it to her and asked if she could switch the site of her work. They approved the change, and she was off to Mauritius, to work with Carl Jones, who was running a project to save the rare birds of Mauritius and Rodrigues.

"I did that for about a year but also tried to get a project together to do something about the plants," she explains. It had become obvious to her that bird species were dying out on Mauritius because their habitat was disappearing, overrun by an exploding human population—only about thirty-five thousand people but one of the densest in the world. She believed that the endangered flora included many very rare plants that were not receiving the attention they deserved: "If species had been that rare in another place, there would have been hoards of people trying to do something to save them."

She wrote to Gren Lucas, chair of the IUCN Species Survival Commission at the Kew Royal Botanic Gardens in England, detailing the number of plants that were going extinct and asking for help in doing something about it. Lucas responded by assisting her to get a small grant from the World Wildlife Fund; later her work developed into a joint project with the Mauritian government. Wendy worked on that project, setting up managed reserves and propagating very rare plants, until she left Mauritius in 1993, and it continued after she left.

Studying the flora of the Mascarene Islands, she was astounded—and depressed—at how rapidly it had been devastated. Unlike in the

Polynesian islands, for instance, explorers landing there had found no inhabitants. Europeans became the first residents.

"The islands had been visited by a few Portuguese and Arab sailors earlier," Wendy says, "but the Dutch took possession of Mauritius in 1598 and made the first real attempt to settle it; the first permanent colonization on Rodrigues wasn't until 1760." By the time the Dutch abandoned Mauritius in 1710, most of the accessible ebony trees—extremely valuable but also slow-growing—had been cut down. "People blamed the Dutch for having wiped out the lowland forest," she explains, "but it probably wasn't all their fault. The French took over the island in 1710 and cut down more forest, before handing the island over to the English in 1812. They completed the job." As far as the most famous extinct animal—the dodo—is concerned, she agrees that it disappeared under the Dutch—but not just because the large, flightless bird had no fear of humans and was easily caught and eaten. The major problem, according to Wendy, was that the Dutch brought pigs with them, which became feral on Mauritius and which continue to cause great problems. "These probably devoured the dodo eggs even more rapidly than the Dutch devoured the dodos," Wendy observes. The dodo (*Raphus cucullatus*) was gone by 1681. Its extinction was the first to enter European consciousness: not the first to occur, but the first to be noted. The last specimens died in European zoos. The Réunion solitaire and the Rodrigues solitaire, the other two species in the family Raphidae, disappeared by 1746 and 1790, respectively.

During her first years on Mauritius, Wendy worked on the nine endemic species of birds that remain on the island, eight of which are threatened. "The top three were the Mauritius kestrel, which is a falcon, the Mauritius parakeet, and the pink pigeon," she explains. "They were supposed to be the world's rarest falcon, rarest parakeet, and rarest pigeon." The program had started with the kestrel, of which there were thought to be only a few pairs left. Efforts had been made to breed kestrels in captivity, but the kestrel rescue program had been plagued by problems since its beginning in 1972. A few days after Wendy arrived, three kestrel eggs taken from a cliffside nest in the Black River gorges hatched out and were named Orange, Pink, and Green. "They all turned out to be males, which was a bit of blow," Wendy recounts. They survived, however, and the next year, three more eggs taken from a nest were hatched out in captivity, yielding two males and a female they named Precious. "The only problem was that the next year we put Precious in with some of the first-year males—even though females usually mature at two years—and it turned out she hated all men and she just beat them up. I think she terrified Pink for life." Precious, like all female kestrels, was much larger than the males, and she was, in Wendy's words, "a real pain."

The following year two more baby females that had fledged were taken into captivity, one of which bred with one of the captive males. "These babies were the first captive-bred birds, and since then there've been hundreds," Wendy says. After the right technique had been found, the birds bred rapidly and were released into the wild.

One of the most interesting discoveries from this program was that though wild kestrel babies were known to survive mostly on endemic day geckos, increasingly rare due to habitat loss, the birds raised in captivity had no parents to teach them to eat geckos. When released, these birds hunted a wider variety of prey, including introduced shrews and small birds. The Mauritius kestrel is still threatened, but with constant management and care, it appears to have a good chance to survive.

What bothers Wendy about the experience is that the Mauritius kestrel had been written off by some people as a hopeless cause, as not worth spending conservation money to save. "Nobody would have predicted that the Mauritius kestrel could start living in an open environment," she says. "And now look at it—I mean, they still live in the forest and they still eat geckos, but they have also learned to adapt to changing conditions."

She tells story after story about the efforts to save other birds. Not knowing exactly how to treat them, project workers had relied on trial and error. The parakeets were also down to only a few individuals, perhaps six birds, of which two were females. Here again, they tried raising them in captivity, but one problem after another resulted in the birds' death. One managed to eat the paint off its cage and succumbed to lead poisoning; another, they suspected, died of salmonella. Still another probably perished because they fed it a diet far richer than it would have had in nature. "The worst time of my life was when those parakeets died," she remembers, "because they kept dying and we knew that there were just a few left in the wild. But when I left, after eight years of working with the parakeets and not really having much luck at all, the first ones bred in captivity." Since then, many more have bred in captivity and have been released into the wild. Although its situation is still precarious, the parakeet, too, appears to have a chance.

The pink pigeon is also still living on Mauritius, after a population decline to about twenty birds, which bred in only one site—paradoxically, a forest plantation of introduced *Cryptomeria* trees. Again, the project workers experimented with different techniques to breed it in captivity. Certain pairs didn't breed, infertile eggs didn't hatch, and some fertile eggs were broken by adult birds. "It just seemed that they were hopeless parents, just totally useless, and the way that Carl Jones, who manages the bird project, got around that was to keep changing the pairs," she explains. "Normally aviculturalists would leave the pairs for quite a while in the hope that they would eventually learn to like each other, but with pink pigeons, if one didn't like the other to begin with, they'd never like each other." They switched pairs and took fertile eggs away from "problem parents" to give them to foster parents—doves. "And once that sort of protocol got under way, we actually began churning out pink pigeons, although there was still the problem of inbreeding."

One point here should not be missed, Wendy emphasizes: "The only reason these birds are coming back is that they're intensively managed." The parakeet, for instance, has people working around the clock to care for it. If the money to support the caretakers runs out, if there is a war or some other crisis on Mauritius, "they're gone then." She refers to the IUCN Red List of categories, which includes a fairly new section: Conservation-Dependent. "The parakeet currently belongs to the Critically Endangered category but if its numbers increase, it could be changed to Conservation-Dependent." It will probably never do better than that.

"On Mauritius they will never, ever, be self-sustaining again," Wendy states firmly. "It is a very artificial world because of the introduced plants and animals, which you can't get rid of. From Dutch times, over the past two centuries, people have introduced black rats (*Rattus rattus*), which are the worst, but we also have feral pigs, deer, cats, mongooses, and others, which all cause incredible damage to the native fauna and flora. And that's not counting the introduced plants—there are now more introduced plant species on Mauritius than there are native."

Still, it was the flora, the nine hundred native plant species, that really drew her interest on Mauritius. And when she first went to Rodrigues in 1982, she discovered another forty-five endemic species of plants on that island. Because only a very old description of the plants was available, a flora by J. G. Baker published in 1877, she found it a difficult process to learn to recognize the native plants. Very few people knew them.

Wendy met an elderly botanist on Mauritius, Dr. Reginald E. Vaughan, whom she considers the father of Mauritian plant ecology because of the depth and breadth of his knowledge of the island's flora. As he was then too old to go into the forest, she collected plants and brought them back for his identification. "And I learned the plants that way," she recalls. "Joseph Guého, Gabriel d'Argent, and Axelle Lamusse also know the native plants, and with them we worked on making an inventory of the native species as well as contributing to the *Flore des Mascareignes,* a flora which is finally replacing Baker's 1877 flora."

Of the native plants of Mauritius, she estimates that thirty species have dwindled to very small populations that have stopped reproducing. "There are ten species on Mauritius and Rodrigues which are down to one or two individuals, which are the real 'living dead.' But there are over two hundred more that have stopped reproducing in the wild, and many of these are on the edge of extinction." Some of these, she adds, have not yet been reduced to tiny populations.

On Rodrigues, she heard the story about Raymond Ahkee, a local schoolteacher who had told his class in 1979 about several extinct plants, including the "café marron" (*Ramosmania heterophylla*), which was purported to have medicinal uses. It had been described in 1874, when it was already rare, and had not been seen since. A child in the class told Ahkee that the plant grew in his garden at home. Skeptical, the teacher investigated: sure enough, *Ramosmania* was there. It took a year to certify the identification, and when Wendy arrived in 1982 she saw it growing by the side of the road, with a newly erected fence around it for protection. Fencing off the rare plant drew the attention of the local people. They began calling it a "magic plant" and claimed it could cure everything from venereal disease to hangovers. Each took a small piece to cure some complaint. When Wendy went out to take a cutting for propagation, she found that the plant had been cut down so far that she decided to wait until it had recovered; however, pieces continued to be taken from the plant, despite the construction of a second fence. Two years later, Wendy took two small cuttings, which were flown within twenty-four hours to Kew. One of those cuttings took and now grows at Kew, but it doesn't set seed, on account of a faulty stigma. The plant on Rodrigues is still alive, with four fences around it at last count. "It's got a guard to keep people away," Wendy relates, "but it's very well known, and people still manage to break in and take some of the plant, although less often than before." In recent years, it has been the subject of an educational program on the island. Wendy points out there has never been any scientific proof that the plant has medicinal qualities.

An even more extreme example of the danger of drawing attention to a rare endemic species is, for Wendy, "the saddest one." In the late 1970s, botanists had shown interest in a rare tree hibiscus (*Hibiscus liliflorus*) growing on top of the highest mountain on Rodrigues. When Gerald Durrell went out to make a film, he discovered that it was being eaten up by goats. Money was given to fence it off from the goats, and forestry department workers were sent out to put up the fence. When the workers realized that they didn't have enough fencing to go around the entire plant, they chopped off one of its large branches to make the hibiscus fit inside the fence. "They didn't quite get it," Wendy laments, shaking her head. "I saw the hibiscus in 1982, just after it had died." She was given new hope, however, when she learned that a cutting taken from the plant had survived and was growing in a Catholic priest's home on the island. "Again, when the plant was fenced, people went up and began taking bits of bark, branches, and also leaving money and putting little candles on the tree," she recalls. "It became another 'magic tree.'" In 1982, when she went up with a forester to take a cutting, only remnants of the hibiscus remained, its fate sealed with wax from the candles placed on it and burnt in its honor. She stood there, staring at the remains, realizing nothing more could be done for it. As she contemplated the dead plant, the forester climbed over the fence and began picking up the change strewn around it. Then she recognized the second tragedy: "People were still throwing money in— these were poor people; they didn't have 'loose change' to spare."

She traces the locals' fascination with the plant to the African heritage of 98 percent of the population on Rodrigues; belief in witchcraft and magic plants and animals is part of the culture. Frequently, she says, special trees were covered in candles and crosses were carved into the bark, or little twigs were taken and pieces of fruit stuck into the holes. It was a situation she didn't see very often on Mauritius, probably because of the different ethnic origins of the Mauritians.

Taking note of the local people's reaction, she began to be careful about paying obvious attention to plants she knew were rare. She thought twice before putting a fence around anything. Instead, she tried to encourage the forestry service to grow more native species and to fence in large areas of native forest.

Wendy and Carl Jones also set up the Mauritius Wildlife Appeal Fund which, among other things, established a wildlife club to further conservation education. She believes that education is extremely important if people's behavior is ever to change.

Wendy thinks that the plants on Mauritius, like the birds, may have a future, even as they slip down to the minimum number required for survival. "I like to think, even when things have gone down to ten individuals, there is still a hope of getting them back," she asserts, almost defiantly. "The other argument is that by having so few individuals, they are hopelessly impoverished and genetically unbalanced, with no hope in the long run; however, I don't believe that. From what I have seen, things can go through a bottleneck and then bounce back. And I think they will have a future as well."

She nevertheless admits that on Mauritius and Rodrigues, some species dropped down very quickly and then didn't bounce back. They went extinct. In the case of *Ramosmania*, she calculates that there is only one individual, and little hope of finding another. "The thing doesn't even produce fruits. Probably the only way we're going to keep it is through cuttings, and the plant will eventually die, unless new techniques are found. At the moment, it is just like a museum piece, although a very important one." And, she emphasizes, it is worth keeping as a museum piece.

She lists other "living dead" and describes the struggles to keep them from disappearing. There's a single palm tree on Mauritius, *Hyophorbe amaricaulis,* which produces male and female flowers at different times but, even with artificial pollination, doesn't produce fertile seeds. It has been grown in vitro, but the seedlings die when planted out of sterile medium. A single *Pandanus pyramidalis* also existed, which she recently learned has died. "We never got it into cultivation, and when I heard that it had died, I felt as if it were someone that had died—there's just no bringing it back."

She would like to see a number of safe places established for each plant species on Mauritius. "And when I say safe places, that means fenced-in, weeded reserves that would be entirely managed." She has surveyed the island and knows well that no pristine forest remains, but some areas are more promising than others.

Wendy does not agree with Norman Myers's statement, in his book *The Sinking Ark*, that limited conservation funding should be spent where it can do the most good. She interprets Myers's triage theory as meaning, "The things that will probably carry on anyway, let's ignore those; let's ignore the things that are really going to cost a lot of money, let them go extinct; let's concentrate on the middle-range things that will make a difference." She remembers the time on Mauritius when they had no money for almost two years and it looked as though they weren't going to be able to keep the bird project going. Then a real estate developer with an interest in birds put up enough money to support the project temporarily, and soon thereafter the Jersey Wildlife Preservation Trust stepped in to ensure that it would continue. Nothing was written off, and in the end their efforts were repaid.

Knowing that it is necessary to establish priorities, Wendy makes several suggestions. On Mauritius they followed the IUCN categories of threat, by giving priority to saving an endangered species before a vulnerable one, for example. Other concerns, too, must be taken into consideration, though: how feasible it is to save a species, what use it has, and what other values it represents. On Mauritius Wendy and her colleagues tried to save and restore habitats, usually by singling out flagship plant species. In that way, other species in the same habitat would also be protected. Again and again, they found that the only way to get funding for work on plants was "to tack it on to the coattails of more sexy-looking birds."

She confides that she was attached not only to the numerous bird and plant projects on Mauritius but to Carl Jones, the man running the bird and bat projects. The house they shared was filled with animals: a monkey named George that regularly escaped and terrorized the neighborhood, a baby endemic fruit bat that had to be fed every three hours for two months, rare birds that needed pampering. Work didn't stop when she went home.

All good things come to an end, though, she says, and the WWF, after funding the project for nine years, decided that it was time for other sources to take it over. At the time, Wendy was finishing her doctoral work on the vegetation of Mauritius at Reading University in England, and an opportunity came up for "a real job" at the world's largest international conservation organization, the IUCN. She says she applied "more or less for the hell of it," and to her great surprise, she was hired.

Wendy is now the plants officer for the IUCN's Species Survival Commission, a position that has made her even more conscious of the high priority generally accorded to projects involving animals, especially large mammals, within conservation groups. "For example, we have a very active African Elephant Specialist Group," she says. "It is amazing all the activity that goes on to conserve elephants—meetings, glossy journals, helicopter surveys, tons of fieldwork, etc. And while I certainly don't want the African elephant to go down the tubes, that's *one* species. With plants we also have specialist groups—for orchids, cacti and succulent plants, pteridophytes—" She stops, then speaks very deliberately and slowly: "But there are some twenty thousand species of orchids, twelve thousand species of cacti, and nine thousand species of ferns—many of them even more endangered than the elephant!"

She laughs, softening the intensity but not the seriousness of her previous comments. She then tells a story to describe the discrepancy in funding accorded the endemic plants and those grown in commercial or agricultural projects. On Rodrigues, she says, her project ran on a ten-thousand-dollar budget. The FAO had a big project on the island to revitalize agriculture. It was valued at thirty-three million ECUs. "They called the manager *Monsieur Trente-trois Millions*—every Rodriguan knew he had thirty-three million ECUs; he had a bigger budget than the entire budget given to Rodrigues from Mauritius." Wendy observed the FAO workers putting up fencing to keep the goats out of agricultural areas, and mentioned in a friendly way to *Monsieur Trente-trois Millions* that she could certainly use some of that fencing for a nature reserve. With never a question, he gave it to her. "It was a drop in the bucket to them," she comments, adding that the head of the program justified the gift as protecting watersheds. "The important point is that we did things through that agricultural program that we could never have achieved through our small conservation grants. I used to dream, wouldn't it be nice to get ten kilometers [six miles] of fencing to put around this little reserve? And then one day it happened."

She also didn't hesitate to mention to the right people on sugarcane plantations that she could use help pulling weeds at her study sites. Laborers who work for the sugar industry that dominates Mauritius cut sugarcane only five months a year; thus surplus labor is generally available the rest of the year. Wendy was also aware that by law, laborers have to be employed year-round. "We then trained the laborers in which plants were native and which plants, such as strawberry guava and privet, we didn't want—and they just pulled out and weeded large areas of the forest that way." Weeding is back-breaking labor, but "sugarcane cutting is as well, so we had a large and motivated labor force to help us on conservation projects."

In the same way, she convinced visiting British and Australian military officials that it would be good training for their helicopter pilots to transport fencing to mountaintops or mist forests where she wanted to fence off endemic plants from pigs and deer.

But there was little she could do to keep out monkeys, a major pest and problem for endemic plants. Not only is Mauritius overrun with wild long-tailed macaques introduced from Java, but the breeding of monkeys for medical research has become a new industry. Most of those exported from the island are captive-bred monkeys, so the wild monkey problem on the island has not improved.

If people's behavior in relation to the environment is to be changed, Wendy suggests, "every trick in the trade" needs to be employed, including enough economic incentives and and other improvements in people's lives that they don't *have* to exhaust the environment. She believes that people are capable of change if they see that it will be helpful to them. In her view, major changes in consumption are particularly necessary: "Europe is bad, the U.S. is worse." She acknowledges, though, that even for her it is not easy. She drives to work rather than take the train because the train ride would take much longer. "Somewhere back in my brain I know I shouldn't be doing this," she admits. "And I worry about living such a comfortable lifestyle in Switzerland, using lots of energy—energy has to come from somewhere."

At the IUCN she brings scientists and researchers, quite often from universities, together with conservationists, quite often from governments; the link will ensure that technical evaluation and expertise will be available to decision makers when they need it. In the process, she has learned a lot more about plants throughout the world, the variety of problems they confront, and conservation in general. "I wish I'd known a lot of what I know now when I was on Mauritius," she reflects. "But I think that the saying 'Think globally, act locally' is of utmost importance. My job is now to try to help people 'acting locally' to do so in the best and most efficient way—and it is a challenge."

Hylobates agilis. Black Gibbon (Freer accession no. 16.47), an early seventeenth-century copy of a painting by I Yüan-chi, the great animal painter who died about A.D. 1064. Courtesy of the Freer Gallery of Art, Smithsonian Institution, Washington, D.C.

II

THE ECHO OF
THEIR SONG

The gibbon . . . was the traditional, purely Chinese
symbol of the unworldly ideals of the poet and the
philosopher, and of the mysterious links between man
and nature. The gibbon initiates man into abstruse
sciences and magic skills, and it is his calls that deepen
the exalted mood of poets and painters on misty mornings and moonlit nights . . .

So important was the gibbon in Chinese art and literature, that he migrated to Japan and Korea together
with the other Chinese literary and artistic motifs,
although [neither] Japan nor Korea ever belonged to
the gibbon's habitat.—R. H. VAN GULIK, *The Gibbon in
China*, 1967

 INTERVIEWS with contemporaries deliver a view of
extinctions through the lens of a microscope—de-
tailed and personal—with the immediate emotion
of direct experience. History delivers a complementary view of extinctions through the lens of a
telescope. We connect in a different way with the
reports of people who died centuries ago, no matter
how feelingly written.

The longest continuously documented history of any civilization
is China's, and one animal that became locally extinct in central China
midway in that history, about A.D. 1000, was the black and white gibbon, *Hylobates agilis*. It played an important role in Chinese poetry
and painting, a role that has lingered on for centuries after the disappearance of the animal itself. The gibbon was a flagship species, representative of many that vanished from central China—the one we
remember, for it was noticed.

This story comes courtesy of a remarkable Dutch diplomat, R. H. van Gulik, probably best known for his Chinese detective stories. When he was stationed in Malaysia, van Gulik had a pet gibbon to which he became deeply attached. When he moved on to China and then to Japan, he traced the two-thousand-year history of the decline and disappearance of gibbons from central and northern China through the poems and paintings of the Chinese scholars and artists who had observed them and been enchanted by their haunting song and graceful movements, and through the Japanese artists who were influenced by them without ever having seen the creature itself. Van Gulik spoke and read Chinese, but he did not think that his literal translations of poems, some of which are presented below, did them justice.

Gibbons are territorial primates, creatures of deep forests and wild mountains, where the male calls, usually at dawn and dusk, to announce his location to the females on his territory and to males holding neighboring territories. The territories are large, and the call is loud and penetrating: it can be heard for up to ten kilometers (six miles). The Chinese found the call of the gibbon in deep forests mysterious, evocative, and haunting.

When van Gulik looked for the first mention of a gibbon in Chinese literature, he could take advantage of Chinese dictionaries of great antiquity. A dictionary compiled around 200 B.C. has a word *nao*, often qualified in Han literature—*nao-yüan* or *yüan-nao*—apparently referring to the gibbon. By the first century A.D., the word *yüan* had clearly taken on the meaning "gibbon," which it kept up to the fourteenth century. During that period, if a Chinese author used the word *yüan*, we can be sure that he was talking about a gibbon. After the gibbon had disappeared from central China and writers no longer had the opportunity to observe directly the animals they were writing about, *yüan* was often applied indiscriminately to macaques and other monkeys. In modern Chinese, gibbons are therefore called *ch'ang-pi-yüan*, "long-armed primates."

In contrast to gibbons, macaques were common. They came into farms and towns to forage and, being often seen and easily caught, they were taught to perform in traveling shows. The Chinese considered macaques inferior to gibbons, which, as they lived in the upper canopy of untouched forest, were rarely seen and nearly impossible to catch. Macaques symbolized trickery, credulity, and tomfoolery. Because gibbons inhabited the forbidding world of wild mountains and valleys, where they were thought to live with magical, imaginary creatures, they symbolized the remote, the mysterious, and the supernatural.

Chuang Tzu, the first Taoist philosopher identifiable as an historical figure, was active toward the end of the fourth century B.C. He wrote a book of parables and fables, which are both charming and wise. Chuang Tzu illustrated one point by describing the behavior of gibbons:

"Has Your Majesty never observed the bounding [gibbons]? If they can reach the tall cedars, the catalpas, or the camphor trees, they will swing and sway from their limbs, frolic and lord it in their midst, and even the famous archers Yi or Peng-Meng could not take accurate aim at them. But when they find themselves among prickly mulberries, brambles, hawthorns, or spiny citrons, they must move with caution, glancing from side to side, quivering and quaking with fear. It is not that their bones and sinews have suddenly become stiff and lost their suppleness. It is simply that the [gibbons] find themselves in a difficult and disadvantageous situation where they cannot exercise their abilities to the full. And now if I should live under a benighted ruler and among traitorous ministers and still hope to escape distress, what hope would there be of doing so?" (This passage was translated from the Chinese by Burton Watson, who used the word *monkey;* but taking into consideration van Gulik's comments on the translation of *yüan,* we have substituted *gibbon.*)

Chuang Tzu had spent enough time in gibbon habitat to observe them as closely as he observed politicians. His parable became a proverb: "If you put a gibbon in a cage, you might as well keep a pig"—not because the gibbon is no longer clever or swift, but because he cannot exercise his abilities. The philosopher feared that the same would happen to him at court.

Later Taoists were more interested in magic and the occult than in Chuang Tzu's logic, sharp humor, and love of paradox. They were convinced that the gibbon was expert at inhaling spiritual fluid, *ch'i.* About 150 B.C., Tung Chung-shu wrote, "The gibbon resembles the macaque, but he is larger, and his color is black. His forearms being long, he lives eight hundred years, because he is expert in controlling his breathing." Some Taoists tried to extend their lives by doing breathing exercises, in which long limbs were thought to be helpful.

The elongated limbs, haunting, melodious cries, and graceful movements of cranes were also thought to indicate that they could inhale ch'i and live to fabulous ages. Cranes became the traditional companions of Taoist saints and hermits; they remain symbols of good luck and longevity in China, Korea, and Japan. Had the gibbon survived in its original range, it might still be assigned a similar role.

Cranes and gibbons became a fixed pair, "yüan-ho," in Chinese art and literature. Their influence extended beyond poetry and painting into music. Players of stringed instruments were told to watch and mimic their graceful movements to improve their fingering technique. One style was associated with wailing gibbons climbing trees—a prolonged vibrato with the left thumb—and another with dancing cranes startled by the breeze—pushing a string out with the right thumb.

Gibbons had a certain dignity. According to the fourth century philosopher Ko Hung, the legendary King Mu of the Chou dynasty made an expedition to the South (ca. 1000 B.C.), where his routed army was transformed. The gentlemen among his troops turned into gibbons or cranes, the "small men" into insects or grains of sand. According to the Confucian ideal, gentlemen were high-minded, superior, and dignified; "small men" were vulgar, immoral, and unpolished. This passage from Ko Hung is cited in almost every subsequent Chinese essay on gibbons. They were gentleman primates, and when a gibbon got very old, a thousand years old, it was thought to change into an old man.

One of the most pervasive gibbon fables, the story of the gibbons forming a chain to drink, first appeared in the historical record in a compilation of Chou texts put together in the first centuries A.D.: "It is hard for a human to get down a high river bank for a drink of water, but thirsty gibbons form a chain and drink from the water below by hanging onto each other's arms" (quotation from the *Kuan-tzû*, as paraphrased from van Gulik 1967, p. 42). This story is repeated in poems of the fifth century and in paintings and carvings. A ninth-century text notes that in the Sui dynasty (A.D. 590–618) the storeroom of the imperial palace was decorated with a ringlike jade carving of gibbons with interlaced arms; such carvings of "gibbon chains" have been reproduced in wood in China and Japan ever since. Van Gulik believed that the story of gibbons forming chains to reach objects below had no basis in reality, but it has had great endurance as myth.

During the Han dynasty the gibbon's habitat, which had extended through central into northern China during the Chou, was progressively eliminated as forests were cut down; however, an island of wilderness remained for many centuries in the midst of increasingly settled landscape—the mountains surrounding the Yangtze gorges in Szechuan and Hupei. Travelers on riverboats beneath the high cliffs could hear the gibbons call in the trees high overhead, and the call of the gibbons in the gorges became a traditional literary symbol in connection with travelers far from home. The cliffs of the gorges, which tower over dangerous rapids, are shrouded in mist, and they inspire awe. The calls of the gibbons, rarely seen but always heard, intensified the mysteriousness of the gorges in the minds of travelers. "In the gorges the calls of the gibbons are extremely clear, mountains and valleys resound with the echo, a deeply sad, continuous wailing. Travelers sang it in the words: 'Sad the calls of the gibbons at the three gorges of Pa-tung; / After three calls in the night, tears wet the traveler's dress'" (Yuan Sung [d. 401], quoted in van Gulik 1967, p. 46).

Gibbons were also kept as pets by the rich and powerful, some of whom became quite attached to them. When the king of Ch'u, Chuang-wang (613–591 B.C.), lost his gibbon, he is said to have had an entire forest cut down in order to find him, in the process no doubt destroying the habitat that had supported a population of gibbons in the first place.

Those wanting gibbons for pets found them very hard to catch in traps. Therefore the method used was to locate a mother with her offspring, kill her with an arrow or poisoned dart, then take the offspring, if it survived the fall of its dead or dying mother from the tree.

Because gibbons were regarded with reverence, however, killing them was seen as a minor sin if not a crime. A certain general Têng Chih, after putting down a rebellion with a punitive expedition about A.D. 250, saw a dark gibbon climbing a mountain and shot it with an arrow. The gibbon's child pulled the arrow out and stuffed leaves into the wound. When the general saw that, he sighed and threw his bow into the river, knowing that he would soon die. The next year another general accused him of plotting a revolt, and he was killed. This story is often quoted in later works, along with numerous cases of archers who, having killed a mother gibbon, either throw away their bows and renounce their trade or are punished by their military commanders.

From the third to the seventh century A.D., Chinese poets, almost all of whom made some mention of the graceful movements and sad calls of the gibbon, often noted when and where gibbons were heard or seen. T'ao Ch'ien, for example, wrote: "Forlornly I dwell in the lonely mountains, / Where the calls of the gibbon are casual but sad" (quoted in van Gulik 1967, p. 52). T'ao noted that his poem was composed in the third month of (A.D.) 416 on the Sun River in Honan Province: a fairly precise time and place.

Gibbons appear in classical Chinese poetry and essays that have been cited as models of their kind ever since (Yü Hsin, seventh century): "The crane's call is of utter loneliness, / The gibbon's song cuts through the entrails"; and "On frosty mornings the cranes cry out, / On autumn nights the gibbons sing" (van Gulik 1967, p. 54).

During the T'ang dynasty (A.D. 618–906), three important essays were written on gibbons in Kiangsu Province. The first was by a Taoist hermit, Wu Yün (d. A.D. 778), a friend of the famous poet Li Po, who had described the gibbon, free of worldly cares, as an example for human imitation. Wu Yün composed his essay on Lu Mountain in Kiangsi Province; the essay can be taken as an indication that black gibbons were abundant there about A.D. 750.

The second essay, by Li Tê-Yü (A.D. 787–847), is set in western Kiangsi. Li had once been a favorite at court, but he wrote his essay after having been banished to a distant provincial post. He compares the aloof gibbons to the quarrelsome macaques and concludes that it is best to follow the gibbon's example and remove oneself from the quarrels and conflicts of the court, where one's fortunes are uncertain.

In the third essay, Liu Tsung-Yüan (773–819) also contrasts the high-minded and well-behaved gibbon with the violent and vulgar macaques, arrogant wastrels, like the louts at court.

The great T'ang poet Li Po (701–762), given the epithet *Chih-hsien*, "the Immortal exiled to the world of men" (van Gulik 1967, p. 59), was a philosophical eclectic with strong Taoist leanings who preferred a bohemian life of wine and song to the high government position that he could easily have had. He grew up in Szechuan, in Ch'ang-ming, not far from the Pa Mountains, where gibbons roamed. He traveled widely and left a record of many gibbon localities of the eighth century. One of his most famous gibbon poems contains the line "A lonely gibbon calls, sitting on the grave mound in the moonlight."

The line was translated into French by the Marquis d'Hervey-St.-Denis, "Ecoutez là-bas, sous les rayons de la lune, écoutez le singe accroupi qui pleure, tout seul, sur les tombeaux."

This was, in turn, rendered from French into German (with some license) by Hans Bethge in his translations of ancient Chinese poetry, *Die Chinesische Flöte* (Leipzig, 1907):

> Seht dort hinab! Im Mondschein auf den Gräbern
> Hockt eine wild-gespenstische Gestalt.
> Ein Affe ist es! Hört ihr, wie sein Heulen
> Hinausgellt in den süssen Duft des Abends?

Bethge's book, and in particular this poem, inspired Gustav Mahler's masterpiece "Das Lied von der Erde," completed in 1908. The gibbon's cultural influence thus stretches from eighth-century China to twentieth-century Europe.

Li Po also left us a famous description of gibbons calling in the gorges:

> In the morning I left the rainbow-clouds of Po-Ti,
> In one day I covered a thousand miles to Chian-ling.
> Before the gibbons on both riverbanks had ceased calling,
> My small boat had already passed ten thousand mountain peaks.
> [van Gulik 1967, p. 60]

And in another unforgettable gibbon image tied to a precise place, he wrote:

> In Ch'iu-pu there are many white gibbons,
> Swirling through the trees as so many snowflakes,
> Pulling their young with them along the thin branches,
> Drinking they play with the reflected moon in the water.
> [van Gulik 1967, p. 61]

Li Po's poems, which provide numerous references to gibbons heard or seen during the eighth century in Honan, Hupei, and Chekiang as well as Kiangsi, allow us to form a minimal estimate of its range at the time.

The gibbon often appeared in classical Chinese literature as an aloof recluse, a model of Taoist withdrawal from worldly cares and deceptions; elsewhere it figured as a supernatural creature with magical powers, a shape-changing deceiver, a were-gibbon or gibbon-ghost, for example, in ninth- and tenth- century T'ang books of fantastic tales.

During the Northern Sung (A.D. 960–1127), a period of strong government, the arts flourished. One Sung master of realism, I Yüan-chi, began to paint roebucks and gibbons. He roamed all over southern Hupei and northern Hunan, and deep into the mountains, to observe gibbons and deer. When he came upon a beautiful scene of trees and rocks, he would study it in detail, to absorb its wild beauty. He would stay with the mountain folk, lingering for months on end. In A.D. 1064 I Yüan-chi was invited to paint screens of gibbons in the Chinese imperial palace. His style was said to be so realistic that after he painted hawks on an government building in Hangchow, the swallows that had nested there moved away.

Paintings of gibbons and roebucks were often done during the Sung, for the gibbon-and-deer combination reflected the same kind of magical mix as did the gibbon-and-crane. The life force ch'i was thought to swell in the antlers and velvet of deer, and deer themselves were believed to be able to find a fungus that prolongs life. They were thus often portrayed eating a fungus, and powdered deer antlers are still in demand as enhancers of male potency and prolongers of life. The deer-and-gibbon paintings of the Sung masters made a deep impression on contemporary poets:

Nothing is more difficult than detailed drawing of living things,
Yet now deer and gibbons are transferred onto this subtle screen.
When these two meet in the wild they first look startled at each other,
Then, in the fragrant breeze, they recognize a common spiritual nature.
By day the deer takes its young to play in the fresh green grass,
At times the gibbon calls its troop to look at their verdant mountains.
Truly lamentable is official life with its dark and devious ways!

This I suddenly realize, every time I pass this screen in the hall.
[Ts'ai Hsian (1012–1067), translation from van Gulik 1967,
p. 79]

None of I Yüan-chi's paintings of deer and gibbons together seem
to have survived, but at least four of his gibbon paintings are still on
display, two in Peking, one in Osaka, and one in London. The almost
photographic accuracy with which the gibbons' slender shape, black
body, and white facial fringe are depicted allows a species identification: *Hylobates agilis.*

The Osaka scroll shows the results of I Yüan-chi's accurate observation of wild gibbons: every gibbon, whether young and old,
walking, sitting, standing, or swinging on a branch, is lifelike, each
gesture precisely realistic. The scroll in the British Museum depicts
two young gibbons, both *H. agilis,* with such accuracy that van Gulik
could estimate their age: the upper one is about three years old, and
the lower about five. The imperial seals indicate that the scroll was
part of the collection of two Sung emperors.

Van Gulik refers to the black gibbon painting in the Freer Gallery
of Art in Washington, D.C., as a "remote copy" of I Yüan-chi's work
that was probably done in the early seventeenth century. The color
painting on silk shows a black gibbon tethered to a horizontal pole
and reaching down for fruit on the ground. "The face is not a real gibbon's face, and the anatomy of the shoulders and the left arm is all
wrong," Gulik remarks. "The setting, however, is quite interesting.
The gibbon is kept on a leash, one end fastened to his neck, the other
to a metal ring that runs along a horizontal pole. This gives the gibbon a fairly wide range, and at present many gibbon-keepers in S.E.
Asia have made such an arrangement in their garden. I may remark in
passing, however, that one should never put a ring round the gibbon's
neck; the leash should be attached to a broad leather belt round his
waist, to prevent him from hurting himself if he has to make an unexpected move" (van Gulik 1967, p. 85).

Impressionistic gibbon paintings have also survived. Impressionism was favored by Ch'an (Zen) monks, whose temples were located
deep in mountain forests. There they had a good chance, during med-
itation at dawn and dusk, of hearing gibbons calling and of actually
seeing wild gibbons. The monk Fa-Ch'ung (ca. 1210–ca. 1275) mas-

tered a style of painting that carried impressionism to its boundary with the abstract: his work features powerful brush strokes, simplicity, directness, and extreme sensitivity.

Fa-Ch'ung grew up in Szechuan, where gibbons were common, and moved east to the Western Lakes, near Hangchow, where monks from Japan came to study; they often returned to Japan with some of his gibbon paintings, which were admired, copied, and preserved. Three scrolls from the collection of the Ashikaga shogun Yoshimitsu (A.D. 1367–1395) can be seen in the Daitoku Temple in Kyoto. One depicts a gibbon mother and child huddling against each other on a branch in cold weather, their fur ruffled against the cold. Although not nearly as detailed as the realistic paintings of I Yüan-Chi, it communicates better the atmosphere of the subject and reflects an important Chinese aesthetic canon: more is on the inside than is expressed on the outside. The gibbons he painted are clearly *H. agilis,* the only species of primate seen in Sung paintings. The other species now found in southwest China and in Vietnam, Laos, Thailand, and Burma, all endangered, did not have ranges far enough north to have been immortalized by poets and artists.

In the thirteenth century Kublai Khan completed the Mongol conquest of China and founded the Yüan dynasty (A.D. 1263–1368). Yüan paintings give indications that gibbons were becoming scarce; some are even depicted with short tails, presumably by artists who had never seen the animal. During the succeeding Ming (A.D. 1368–1644) and Ch'ing (A.D. 1644–1912) dynasties, gibbons were mentioned in poems, depicted in paintings, and kept from time to time as pets imported from the south to the imperial palace, but they were no longer observed living in the wild in central China.

In reflecting on two thousand years of gibbons in Chinese culture, van Gulik wrote (1967, p. 96): "I find it difficult to decide . . . where lies the gibbon's greater glory. In the exquisite ancient Chinese poems and essays recited to this very day, and in the fascinating antique paintings now in museums and art galleries? Or perhaps rather in the clumsy grave-markers in the corners of neglected gardens, bearing pet-names laboriously lettered in a small child's hand, and soon washed away by the tropical rain?" (van Gulik 1967, p. 96).

Van Gulik plotted on a map of China all mentions of gibbons that were associated with a reliable date and place. During the first millennium A.D. the range of the gibbon extended over three-quarters of China, with a northern frontier at about 35° north latitude, where winters can be severe and gibbons were frequently seen in snow. The locations plotted on the map are clustered around the scenic and historical sites visited by poets, monks, and scholars. Gibbons certainly also lived at sites in between. In particular, they must have been in southern China, which was not yet developed in T'ang times. When the Chinese explored those areas in the seventeenth and eighteenth centuries, they sighted gibbons in many places in Fukien and Kwantung. By that time only a few isolated populations remained in central China. In the twentieth century, war, human famine, industrialization, progressive deforestation, and the development of railways pushed the gibbons toward the southwest, where in the 1960s gibbons were still reported in Yunnan and southern Kwangsi. By then they had long been extinct in central China.

It is quite common for human cultures to associate particular qualities with certain animals, which feature an astonishing variety of roles: Joe Camel is cool, Mickey Mouse is clever and funny, the Lion King is courageous and noble. Hawaiian sharks were seen as shape-changing demigods who, like the gibbons in ancient China, could transform themselves into human form. Kwakiutl totem poles were prominently decorated with bears, eagles, and killer whales. The list could be extended through all the cultures of the world.

One thing is distinctive about the gibbon in China and its role in Chinese culture. It became locally extinct right in the midst of things, and the people who cherished it, the educated cultural elite—courtiers, bureaucrats, and monks—understood why it was going extinct even as the extinction was taking place: because the forests were being cut down for firewood by an expanding peasant population.

Reports were written, edicts were drafted, orders came down from on high: the forests were not to be cut down; gibbons were to be protected—but the edicts had little effect. The laws did not change the constraints on the life of peasants. They still had babies, often more babies than necessary to maintain the population, and the expanding population had to cook food on fires fed by wood from the forests.

Furthermore, the authority of the central government was not continuous. Dynasties fell, civil wars raged, and sometimes weak governments prevailed for many years, particularly at the periphery of the empire, at the edge of the forest. Even if a strong central authority could have figured out a way to preserve the forests, that authority would not have persisted for more than a few centuries. The process of extinction lasted two millennia, and the destruction of the forests proceeded like a ratchet: each time it clicked forward, it did not slide back.

In some ways little has changed. We deplore extinctions and understand that one root cause is too many people, who consume too much. We have still not found a way to live in peace and economic harmony with our fellow human beings while sustaining the natural world we value. And now, as then, it is unlikely that a single peasant farmer would trade the life of a starving child for that of a gibbon.

In another way we are different. Whereas the ancient Chinese, for whom extinctions were as yet a new idea that they had not yet fully digested, lived through the local extinction of the gibbons, we can look back at their experience, compare it with what is now going on, and draw conclusions. We have the benefit of a perspective that they did not. The history of the gibbon in China shows us that when we went through all this before, it did not turn out particularly well—neither for the gibbons, which disappeared, nor for the people who could no longer draw aesthetic and spiritual inspiration from them.

12

WATCHING, FROM THE EDGE OF EXTINCTION

WHEN WE STARTED THIS PROJECT, we had few ideas about the conclusions to which it might lead us. We revised our opinions and expectations as we traveled and interviewed people, many of whom described experiences we had never contemplated. We were surprised at the themes that were repeated, at connections we had not seen earlier, at the number of simple mistakes made, sometimes at great cost, to endangered species. And we were impressed by the complexity of the environment and human interactions with it.

We had underestimated the high drama associated with extinctions. We had not expected to find so much corruption and political maneuvering connected with what we thought would be scientific and cultural problems. Nor had we anticipated the huge sums of money being contended for in the fight to save endangered species—or spent in the name of development. We had not thought about involvement of intimate friends of scientists and conservation workers or their exclusion of "outsiders" from conservation projects. We listened, asked questions, and tried to sort out fact from fiction. It was not always easy. We recognize that many stories remain to be told and that some points of view are missing, but we have reported the stories of the endangered species in these chapters as accurately as possible.

In addition to gaining disillusioning insights into conservation biology and human nature, we met people who were a real inspiration. Ismail Mohamed Oban, our driver and guide in Africa, made a lasting impression. A survivor of Idi Amin's reign of terror in Uganda, Mohamed has the kind of street sense we lacked. He is also a humble and thoughtful man. Above his sun visor in the Land Rover he kept a small notebook in which he wrote the names of new animals he observed, and he observed them as keenly as a scientist. Trained as a mechanic—a valuable background for anyone driving long distances in Africa—he took apart the broken transmission on our vehicle on a lonely road south of Nairobi and, after hitching a ride into the city for a spare part, managed to repair it so we could continue our trip into the Serengeti, where he proudly added cheetahs to the list in his notebook. He was self-reliant, unassuming, and modest. Asked what he would do with the money he earned, he replied it would go toward schooling for his son. He didn't need much money, he added—didn't need new clothes, could grow most of the food the family needed, didn't drink or smoke. "If people have a little money," he told us, "they first might buy some sugar for their tea. If they have a little more, they might buy some salt for their food. And if they have more, well, some soap would be nice." He could teach us all a few things about consumption.

There are many perspectives on extinction. One view that should not be overlooked is that not all extinctions are notable or important. At the other end of the spectrum is the comparison of extinctions with a holocaust in which animals and plants are the victims. We present both in this closing chapter, for they frame the boundaries of the reactions we encountered.

Not all extinctions are important. The late autumn sun glistens on the surface of Lake Constance as Dr. Hans-Joachim Elster reflects on the nearly seventy years he has studied the lake, at which he looks daily from his home above the German shore. The main focus of his early interest in the lake, a copepod on which he wrote a monograph in 1932, disappeared many years ago. Few residents on the shores of the lake noticed its disappearance, but it was among the most striking events of his career. A handful of limnologists found it an interesting phenomenon. Dr. Elster states explicitly that its extinction meant little. It was, in fact, probably like most organisms that become extinct: no one notices, and the extinction has no consequences large enough to raise questions about what happened.

Born in 1908 in what would become the German Democratic Republic (GDR), Dr. Elster studied at Leipzig and Munich with some of the great names in German biology. His promise was recognized early: at the age of twenty-three he was asked to found a limnology institute on Lake Constance, known to German speakers as the *Bodensee.* Before World War II the research station was called the Langenauer Institute and was part of the Kaiser Wilhelm Institute, the predecessor to the current Max Planck Society.

To ascertain how they could increase the production of fish in Lake Constance, Dr. Elster and his team investigated the copepod *Heterocope borealis,* a small crustacean on which the economically important fish in the lake fed. The scientists working on Lake Constance admired *Heterocope* for its graceful jumping motion, its relatively large size (a few millimeters), and, when it was well fed, its beautiful blue color, which turned to red in the stomachs of the fish that ate it.

In the fall, *Heterocope* laid its eggs in the open waters of the lake, which is 200–350 meters (650–1,050 feet) deep. The eggs sank to the bottom and rested on the sediments until spring, when the larvae emerged in huge numbers and soon moved into the upper waters. Then their impressive daily migrations began, the copepods rising to the surface at night and swimming 50 meters (165 feet) or more into the depths during the day. One kind of whitefish fed almost exclusively on *Heterocope* and followed the copepods in their vertical migration. Feeding at the surface during the night, the whitefish were caught by fishermen, in boats whose lights in those years typified summer evenings on Lake Constance.

Heterocope was abundant and important in the lake ecosystem, and its extinction in the 1960s came as a surprise. The intensification of agriculture around the lake washed fertilizer into its waters, triggering algae blooms on the surface. When the algae died, they sank into deep water, where their decomposition used up the oxygen, thus making it difficult for *Heterocope*'s eggs to survive the winter. The limnologists could see from the black color of reduced iron on the sediment surface that the oxygen was nearly gone from the deepest water. To cap things off, a predatory copepod from the shallows moved into open water in spring, when the *Heterocope* larvae hatched, and gobbled them up.

When *Heterocope* went extinct, Dr. Elster says, the whitefish that had fed on it did not suffer—they prospered. The water flea, *Daphnia*, including a species new to the lake, increased its numbers enormously. The whitefish, which had few natural enemies, ate the *Daphnia* and grew more rapidly. The authorities adjusted the quotas for the fishery upward; the fishermen's yield increased threefold.

Heterocope, a striking organism formerly present in huge numbers, is now extinct. Does its extinction mean anything? Dr. Elster notes that other *Heterocope* species survive in other lakes, and its extinction in Lake Constance does not appear to have set off a chain reaction of other extinctions. Instead, new species have invaded the lake. Only the limnologists had noticed that *Heterocope* was ever there. Its disappearance was an interesting intellectual puzzle but did not elicit profound emotions.

Dr. Elster tempers his dry conclusions with some concern about the future. Now that the lake is becoming cleaner again and *Daphnia*, with less algae to eat, is declining in numbers, the whitefish may not be able to maintain their former abundance in the absence of *Heterocope*.

He is also philosophical. Extinctions make a difference, he says, for we are social beings and other species are part of our social environment. He quotes Albert Schweitzer: because we live not in the *Umwelt* (the environment that surrounds us) but in the *Mitwelt* (the world that accompanies us), we have no right to dominate other life.

Some extinctions are a holocaust. Another scientist we met had witnessed the extinction of hundreds of species of fish in Lake Victoria that was caused by the Nile perch. He is not one whom we mentioned in the chapter on the Nile perch, but he was nevertheless deeply moved by what he saw happen. He confided to us that he had from the time he was very young suffered from bouts of depression that he could neither understand nor justify. He felt guilty about this depression, and he became further depressed when he thought about his father, who, having lived through the horrors of the Bergen-Belsen concentration camp during World War II, had much greater reason to be depressed. Only when he experienced firsthand the extinctions of hundreds of haplochromine species in Lake Victoria was he finally able to justify his depression. He felt then that he, too, had witnessed a holocaust.

We were speechless. The only possible responses were existential tears or existential laughter. We chose laughter, and laughed with him, shaking our heads and crying inside.

Having reflected deeply on what we learned as we wrote this book, we offer the following conclusions:

The long-term solution is to have fewer people—and for those people to consume less. Plants and animals would not be going extinct at the rate they are if human consumption were held within certain bounds. Virtually no measures can be taken—no technical fixes found—that would fully repair the environmental damage done by overpopulation. When environmental damage drives species to extinction, it acts inexorably, moving us, sometimes slowly, sometimes rapidly, in only one direction: toward an impoverished living world.

The developing world has a population problem, but the impact of the human population is much worse in the developed world. We see the species vanishing now from the tropics, but that is only because we do not have the records or the history to show us how many species have already vanished from developed countries. It is not fair for us, having already exploited our own environment, to ask others not to exploit theirs unless we can first bring our rate of consumption under control. The United States, for example, with 5 percent of the world's population, uses 24 percent of the world's energy and produces 19 percent of the world's garbage. As Aliki Panou put it, at the Earth Summit conference in Rio de Janeiro in 1992, citizens of the Third World were told that they couldn't each have one car because each resident of the developed world has two already.

Money breeds corruption. As soon as money is made available to protect endangered species, people who were not otherwise interested in the species become interested in the money; they invade the institutions set up to protect plants and animals. This opportunism can result in conservation organizations whose budgets go mostly to maintaining central offices and staffs with a pleasant life style rather than to projects that save species in the field. Conservation has become a trendy career, which seems to offer the opportunity to do good while living well. We have heard research scientists describe some conservationists they have encountered as "eco-mimics," people who appear to be ecologists but do little except benefit financially or socially from that role. The more people we spoke with, the more we realized that this is a global phenomenon, and that serious scientists and dedicated conservationists suffer the consequences—as do the species that remain unprotected.

When Switzerland made grants available in the early 1990s for people to work on environmental and conservation problems, hundreds of grant-seekers simply relabeled the work they had been doing to make it appear essential to conservation. This ploy is not specific to Switzerland, it is routine behavior. In the business sector—worldwide—people are restructuring to take advantage of the current opportunities.

Human motivations are often mixed—and muddled. Perhaps we should expect selfishness and be pleasantly surprised by idealism, even from those who work with endangered species. But the public has a right to a high ethical standard from conservationists who themselves employ ethical arguments to tell other people what to do. When the public realizes that a conservationist, while claiming to save a species, is not helping it or even is actually contributing to its destruction, the damage to the image of conservation compounds the damage to the species. This we saw most clearly in the stories of the Hawaiian crow, the monk seals in Greece, and the wild dogs in the Serengeti.

Even conservationists deeply concerned with ethics sometimes portray their work in a way they know may be more appealing to the public—or to the government. To get permission for their work on Lake Victoria, Dutch researchers told the government they were doing research on the food of the Nile perch, not on the disappearance of the haplochromines. It was, of course, essentially the same thing. Such relabeling extends to social appeals, such as making the Barton Springs salamander the symbol of the city of Austin to help stimulate interest in its conservation. This approach is not necessarily misleading, but we should reflect carefully on how scientists, conservationists, politicians, or developers present the facts. Truth is fragile, and reality is subject to alternative interpretations, especially in fields like conservation that are highly charged with social conflicts.

Scientists probably are in the minority among conservationists, and their work is, in general, commendable. Nevertheless, some of those working in conservation do so mainly for personal gain, which for a scientist has as much to do with publications and recognition as with money. Cynthia Salley saw this clearly when scientists repeatedly disturbed the breeding sites of Hawaiian crows to gather data to publish papers that would help the scientist but not the crow. Her proposed solution—that scientists could come onto her property to protect the crow, but that they would not be allowed to publish the results of their work—hit a sensitive nerve.

Competent people are in short supply. Most endangered species do not meet with the best that humanity has to offer. Their fates are often determined by people who may be well-meaning but unskilled or incompetent. They may also be foolish, lazy, or even corrupt; their actions can result in catastrophe. Even competent but poorly informed people can love endangered species to death. That was particularly clear in the cases of the Hawaiian crow and Britain's large blue butterfly. There was an element of it also in the stories of the Hawaiian land snails and the wild dogs of the Serengeti, where the emotions may not have been so pure.

A corollary is that a few individuals can have enormous power over the fate of entire species, for good or for ill. The people who make the biggest difference are those who decide in the first place to get involved. Support for work on endangered species, sometimes meager, ensues. Resources follow leadership, and both are critical. In the case of the Barton Springs salamander, a whole city was led by a few: their numbers grew until a citywide resolution was passed, and until the U.S. government and the country's highest courts could no longer ignore the concern voiced by the people of Austin, Texas.

Supply and demand are always at work, whether you see them or not. If no demand existed for more resorts, hotels would not be built on precious natural habitats; if the demand for the meat of forest animals were not so strong, bushmeat restaurants in Abidjan would not be doing a booming business; if no one bought rhino horn or leopard-skin coats, the animals that produced them would not be killed.

Similar to the invisible hand of supply and demand, at least in its invisibility and relentlessness, is the chain reaction set in motion when species are removed from the ecosystem or introduced into it. This we saw most clearly in the story of the Nile perch; dumping a bucket of fish into a lake had consequences almost unimaginable in their scope and complexity.

Most scientists do not make effective lobbyists. Scientists have been shaped to a different standard than politicians, promoters, or developers. If they know of an alternative interpretation, they feel obliged to mention it. If they are not 100 percent certain, they say so. Their honesty and objectivity may confuse the public and make a clear choice difficult. Politicians are confronted, on the one hand, with lobbyists for industry who are trained to exaggerate their position and to speak with certainty, and on the other hand by scientists who qualify their statements and go into detail. Either some scientists should learn how to lobby or trained lobbyists should speak for the conservation movement.

Some extinctions take with them a piece of human history and culture. We have emerged, from periods in which we were hunter-gatherers and rural farmers, when we had special relations with nature, as primarily an urban civilization; our culture incorporated symbols and stories built around other species, some of which are now extinct or endangered. When a species that had accompanied us along the way vanishes, it carries with it part of our own history and some of the meanings that we expressed in art and language through natural symbols. One clear example is seen in the story of the gibbon.

There are political and financial aspects to most of these stories, although some people may consider extinctions to be of only scientific interest. But what about the cultural and aesthetic angles? Poets and artists have something to say, too, but either they have not spoken loudly enough or their eloquence has been lost on those who could have made a difference. Emotions can be just as important as facts when it comes to preserving our surroundings. Such decisions should not be relegated to scientists, lawyers, or politicians. We are all affected; we should all speak out.

Some people believe species have rights, including the right to exist. The people we interviewed had varying opinions about this. Christophe Boesch, Aliki Panou, and David Hillis all believe to some extent that other species do have the right to exist, whereas Cynthia Salley and Heribert Hofer expressed the view that the very concept of having a right is an artifact transferred from the human world, not something that exists in nature independent of humans.

There is another way to look at this: all species share the planet. We humans happen to have evolved language and technology and to be in control at present, but that does not mean we have an exclusive or "natural" right to dominate. This is what Dr. Elster was getting at when he quoted Albert Schweitzer: that we live within the world— it and we are fellow travelers. If any other species on the planet had attained our cultural and technological capabilities, it too would probably display a tendency toward short-sighted selfishness. But we, the most powerful, have the responsibility of the powerful to control ourselves—especially our consumption, reproduction, and ability to destroy. Until we do, we are nothing more than unusually intelligent animals with unusually sophisticated tools, animals governed, with regard to important questions that determine the fate of most living things on earth, more by instincts than by reason.

Some things are better left alone. Many people, especially the active, the engaged, and the ambitious, have a hard time with the idea that it might be best to do nothing at all. Their automatic reaction to a problem is to take action to solve it. Sometimes taking action makes the original problem worse; sometimes it creates entirely new problems. Consider the stories of the Hawaiian crow, the wild dogs of the Serengeti, and the Nile perch. Natural systems are complex. The patterns we see may not have the causes that we at first suppose. Our ability to predict the consequences of intervention is limited. Yet in the situations recounted here, an attitude of self-critical modesty was often in short supply; activism was the predominant response.

This approach is far removed from that advocated by Chuang Tzu, a Taoist writer and philosopher who lived about 300 B.C. The following text, written by that wise man many centuries ago and translated by another in modern times, reflects a simple but deep insight: "The wise man, then, when he must govern, knows how to do nothing. . . . he will govern others without hurting them. Let him keep the deep drives in his own guts from going into action. Let him keep still" (Merton 1965, p. 71).

We do not mean to imply that nothing should be done, but it is easy to do too much too quickly without thinking about alternatives or consequences.

Constant management is the only way to preserve some species. The world has been irreversibly changed; there is hardly any place where nature remains truly "natural." Humans have so altered their environment that many things that once existed comfortably, although perhaps they never flourished, can no longer survive in their native habitats. Jeremy Thomas recognizes that this is true of the large blue butterfly in Britain, Wendy Strahm sees it with hundreds of endangered plants on Mauritius and Rodrigues, Mike Hadfield fights to keep more introduced snails from eating up the endemics on the Pacific Islands. Do we work to remove animals and plants from the endangered list just to put them on the conservation-dependent list? We can, but it is costly and uncertain.

Species that are dependent on management will survive only as long as funding and social stability provide for their future. Management rapidly disappears during wars and economic hard times, leaving animals vulnerable. This happened to the gibbons in China 1,800 years ago and, much more recently, to the mountain gorillas in Rwanda and Zaire. When civil war hit Zaire and food became scarce for both animals and the zookeepers, the keepers ended up eating their charges. Management can also disappear rapidly when the government changes priorities, and when programs reliant on "soft money" lose their support. The first requirement for long-term conservation may well be enough social justice and economic well-being to head off wars and revolutions. It is therefore extremely important that we determine our priorities and hold to them in the long term.

What are our priorities? Do we value nature enough to make some changes? Do we believe a diversity of species adds significant value to our lives? Can we get along without more beach resorts and hotels on the breeding grounds of monk seals and turtles? If we have boxes of beautiful shells and stuffed exotic birds in museums, do we really need to see them in the wild?

If you answer no to these questions, then at least you have made a conscious decision not to do anything and you carry the responsibility for that decision. But if it does matter to you and you don't do anything about it, then you let those who profit from their exploitation of nature become rich while stealing away the irreplaceable beauty and diversity of the Earth from you, and from your children and grandchildren. Ignorance may be no excuse, but informed apathy is worse.

Early on in this project, we spoke with an economist who generally moves in international banking circles but who shares our interests in literature, history, and natural history. He listened intently as we listed possible chapters for a book about endangered species. "I think your portfolio is too large," he said after we concluded. It was a simple statement from an economist's perspective, honest and direct. We can't save everything, he was saying, we need to set priorities, decide where to invest.

His comment came back to us when we asked Storrs Olson how he would reconcile two different forces—people who don't care about extinction because they have to eat, and people who care because they do have enough to eat and want to live in an aesthetically pleasing world. Storrs responded: "I don't know, I'm not an economist."

Conservation of species and resources has to make practical and financial sense to be understood and accepted by most people. Jim Jacobi argues for healthy watersheds that will also save the indigenous local flora; Barbara Mahler recognizes how much more expensive it is to treat surface water than it is if you have to pump water out of the ground. As Nile perch shrink in size and the indigenous species in Lake Victoria disappear, it is becoming clear that the introduction of the Nile perch may not result after all in the sustainable fishery that so many people touted.

Some scientists try to put a dollar value on the services performed by ecosystems, by calculating how much it would cost to replace them with technological fixes. They argue that if such services had to be paid for in proportion to their contribution to the global economy, the price structure of the world's markets would be very different. Many businesses whose profitability depends on unsustainable exploitation would disappear. But are most people ready to acknowledge the economic value of ecosystems—and to pay the price necessary to protect them?

The dollar-value reasoning of some scientists and economists is appropriate on one level; at other levels, some values inherent in natural ecosystems cannot be expressed financially. Think of all the things in your life that could not be replaced by money. They range from family heirlooms with strictly personal meaning to good health and loved ones themselves. So it is with our reactions to very important parts of nature. You can measure the difference with economic valuation by comparing how you react to the artificial nature marketed in theme parks with your reactions to real wilderness. The difference, for many of us, is enormous, for there is no substitute for the real thing, no restitution for things stolen that cannot be replaced.

Shortly before we finished this book, Bev was invited by a Swiss friend to go hiking in the mountains. It was early June, and the wildflowers were beginning to bloom. Some things cannot wait. She turned off her computer and went. During that extraordinary succession of warm days in the mountains, Nature put on an amazing display: hundreds of the endangered lady's slipper (*Frauenschuh*) orchids like those Hedwige Boesch's father had warned her not to pick, fields of fragrant wild narcissus, mixed with gentians and anemones, crocuses and soldanellas, which, as the high snowfields melted, came up for their moment in the sun. Farmers in the valleys, when they did the haying, laid up not only grass but also orchids, buttercups, and marguerites for the cows' winter food.

One night Bev sat among a small group of Swiss who had grown up in the mountains, gone to ski school together, and then left to travel and live around the world. Now, in their sixties and seventies, they were back, enjoying being among old friends, speaking local dialects. They conversed passionately and movingly on that warm June evening. Banking, politics, sports, and favorite vacation spots were never mentioned. Nor was music or literature or art, although those were all subjects they could have discussed with ease. No, they spoke exuberantly about the wildflowers they had just walked through, comparing the names in various languages. They whispered the location of especially beautiful paths, gave directions to woods where a particular rare flower could be found, noted thoughtfully the sighting of a bird, probably in migration. They exchanged information as earnestly as any financial adviser would offer counsel—information that indeed seemed as precious as gold. It was a privilege to be part of the conversation.

Bev did not tell them that although she found the Swiss wildflowers lovely and the mountains breathtaking, the forests and valleys seemed terribly empty. She did not remind them that the last bear in

Cypripedium calceolus. This is the lady's slipper, *Frauenschuh,* now highly endangered and rare, which Hedwige Boesch's father warned her never to pick. Photograph by Beverly Stearns.

Switzerland had been shot in 1904, that for decades the Rhine has been so polluted that the salmon have not been able to reach Basel, or that acid rain has devastated most Swiss forests. She did not tell them these things, for she knew that they were aware of them. Perhaps that was why their passion for the wildflowers was so great.

The Swiss, at the heart of Europe, in the midst of success, with one of the highest per capita incomes in the world, know the full consequences of the economic development sought by those who do not have it. Some Swiss have great wealth, practically none are poor, but their interest in the outdoors—hiking, skiing, mountain climbing, for instance—is nearly universal. However, what is left of nature in a land that has become rich enough to give its people economic security? The bears, wolves, and river otters that lived there in the early twentieth century are gone, not to mention the cave lions, mammoths, and woolly rhinoceroses of the Ice Age. Nature reserves in local communities are now defined on the basis of beetles and flowers. We do not denigrate the beetles and flowers; we contrast them with the bears, wolves, and otters simply to make clear how much has disappeared by the time people have had enough success in their pursuit of financial security to pause and look around.

Perhaps the rest of us should stop and look around before we achieve complete "success." There may still be time to protect some of the natural world, and not just watch, from the edge of extinction.

Afterword

In September 1999, shortly before dawn, we drove through pristine Hawaiian cloud forest high on the slopes of Mauna Loa to where Cynthia Salley told us we might see the last two 'Alala in the wild. She said there might be a third, but it had not been seen for weeks. When we talked with her in 1997, there had still been twelve wild Hawaiian crows. Just as the sky began to lighten, as we approached the open-air aviary where the captive-reared 'Alala were kept, we heard the call of the wild birds. We stopped, turned off the motor, and quickly got out of the four-wheel-drive vehicle. The birds soon flew on up the slope, perching high in the tops of native koa trees; their calls were unlike those of familiar crows. We drove on to the aviary, a huge wire-enclosed cage. With binoculars we could see the two wild 'Alala silhouetted against the sky, in two old koa trees rising high over the aviary. They were an old pair, past the age of reproduction, calling to each other in the morning light. Perhaps they were also calling to the young inside the aviary. As their clear, plaintive calls went on and on, we stood silently and listened. Inside the aviary, the hand-raised juveniles squawked and jumped around, expecting their morning feeding from people employed to take care of them. In the solitude of the Hawaiian forest, we felt privileged and sad. We were standing in one of only a few patches of native Hawaiian forest, watching the last two wild birds of a species.

A captive breeding program has been fairly successful in hatching and raising the 'Alala until they can be released from the outdoor aviary, but the young birds have had no natural role models and have developed no street smarts. While wild birds stop calling by about 7:30 A.M., the released birds call well into the morning. So far, all the released birds have disappeared, many of them eaten by the endangered *I'o*, the native Hawaiian Hawk.

Shortly after leaving the aviary we saw an I'o bathing in a puddle on the road. It flew into a tree nearby and watched us. An hour later we saw a flock of endangered *Akepas* moving through the crowns of small O'hia trees, the males like bright orange flames.

In the spring of 2000, Cynthia Salley sold most of her ranch to the federal government to be managed by the U.S. Fish and Wildlife Service as a reserve for endangered species. It was no longer financially possible for the ranch to continue to exist under the restrictions imposed on the use of land on which endangered species live.

Let us now praise famous men—
and beasts and plants.

Let us remember the 'Alala
that flew toward the sound of the guns
and delighted the girl who saw them:
her private amakua,
a protecting spirit.

Let us remember the hunting dogs,
strong to bring down zebras and wildebeest,
licking their pups and watching the moon rise,
ears up, noses to the wind,
on the short-grass plains,
east of the Moru Kopjes.

Let us remember the monk seals,
seeking peace deep in their caves,
beyond tourists and motors,
the ancient companions of the Nereids,
on the shore of the sea of Odysseus.

Let us remember the call of the gibbon,
deep in the forest at dawn and dusk,
or high above the boats on the Yangtze gorge,
and the monks, meditating,
and the poet-administrators
traveling to their distant posts,
for whom the sound evoked
all the wildness
and all the mystery of Nature,
untamed, beyond.

Let us also remember
the unarticulated traces
of those which disappeared,
in their hundreds and thousands,
before anyone named and described them,
without witness, abstract,
but of as much value to themselves
as we to us.

What have we lost?
And why did it go?

THE TROPICAL
BIOLOGY
ASSOCIATION

BECAUSE the future of the world's plants and animals depends on the knowledge and competence of people in key positions, the Tropical Biology Association (TBA) was founded in Switzerland in 1990 to help train a corps of experienced biologists to fill those positions. The Organization for Tropical Studies, founded in the United States, is similar; it conducts courses such as the one in Costa Rica that Steve attended, first as a graduate student in 1972 and later as a teacher.

Most of the world's species—including many endangered ones—live in the tropics. Because the people who live there are often unable to obtain the education and support necessary to study the indigenous plants and animals, the TBA was set up to instruct students from tropical countries and from Europe; the courses are held in the tropics and are taught by teachers from all over the world. The tuition costs of students from the tropics are partially subsidized by European universities, which pay membership fees to send their own students to the program. Twenty-seven universities or departments are members. The results have been gratifying: 317 students from thirty-four countries in Europe, Africa, and Southeast Asia have been trained in fifteen courses held in Kibale Forest in Uganda, Hell's Gate National Park in Kenya, and Danum Valley in Malaysia.

The program has been supported by a generous Darwin Initiative grant from the British government, as well as Swiss, Dutch, and E.U. grants during the difficult early years.

A TBA Foundation has been set up to receive contributions toward an endowment that would make it possible for the association to become self-sustaining.

We hope that graduates of TBA courses will become good scientists and educated guardians of biodiversity thanks to cross-cultural insights offered by fellow students from other lands. At Kibale Forest in Uganda we listened to African and European students discuss the extinction of a species, what it means scientifically and economically, whether it mattered, how it might have been avoided. In the background, colobus monkeys leaped among the giant trees, and sunbirds fed from bright flowers. The students' interest and excitement was clear: the teachers had an eager audience for their lectures, readings, statistics, and computer programs.

If you wish to support the goals of the TBA or would like to receive more information on its activities, the address is

Tropical Biology Association
Department of Zoology
Downing Street
Cambridge University
Cambridge CB2 3EJ
England

BIBLIOGRAPHY

General

Conservation Biology, journal of the Society for Conservation Biology, Blackwell Science, Cambridge, Mass.

Diamond, J. 1991. *The rise and fall of the third chimpanzee.* London: Vintage.

Groombridge, B., ed. 1993. *1994 IUCN red list of threatened animals.* Gland, Switzerland: World Conservation Union.

Lawton, J. H., and R. M. May, eds. 1995. *Extinction rates.* Oxford: Oxford University Press.

Primack, R. B. 1995. *A primer of conservation biology.* Sunderland, Mass.: Sinauer Press.

Reaka-Kudla, M., D. E. Wilson and E. O. Wilson, eds. 1997. *Biodiversity II.* Washington, D.C.: National Academy Press.

Preface

May, R. M. 1988. How many species are there on earth? *Science* 241:1441–1449.

Nott, M. P., E. Rodgers, and S. Pimm. 1995. Modern extinctions in the kilo-death range. *Current Biology* 5:14–17.

Chapter 1: Extinctions in Perspective

Berger, A. J. *Bird life in Hawaii.* Honolulu, Hawaii: Island Heritage.
———. 1972. *Hawaiian bird life.* Honolulu: University of Hawaii Press.

Freed, L. A., S. Conant, and R. C. Fleischer. 1987. Evolutionary ecology and radiation of Hawaiian passerine birds. *Trends in Ecology and Evolution* 2:196–203.

James, H. F., and S. L. Olson. 1991. *Descriptions of thirty-two new species of birds from the Hawaiian Islands.* Ornithological Monographs nos. 45 and 46. Washington, D.C.: American Ornithologists' Union.

Milberg, P., and T. Tyberg. 1993. Naïve birds and noble savages—A review of man-caused prehistoric extinction of island birds. *Ecography* 16 (3).

Neal, M. C. 1965. *In gardens of Hawaii.* Honolulu, Hawaii: Bishop Museum Press.

Olson, S. L., and H. F. James. 1984. The role of Polynesians in the extinction of the avifauna of the Hawaiian Islands. In P. S. Martin and R. G. Klein, eds., *Quaternary extinctions: A prehistoric revolution.* Tucson: University of Arizona Press.

Pimm, S. L., M. P. Moulton, and L. J. Justice. 1994. Bird extinctions in the central Pacific. *Phil. Trans. Roy. Soc. Lond. B* 344:27–33.

Steadman, D. W. 1995. Prehistoric extinctions of Pacific Island birds: Biodiversity meets zooarchaeology. *Science* 267:1123–1131.

Van Neck, J., and W. Warwijck. 1601. *De tweede schipvaart naar Oost-Indië, 1598–1600.* Amsterdam: Linschoten Vereeniging.

Chapter 2: Empty Shells

Clarke, B., J. Murray, and M. S. Johnson. 1984. The extinction of endemic species by a program of biological control. *Pacific Science* 38:97–104.

Hadfield, M. G. 1986. Extinction in Hawaiian Achatenelline snails. *Malacologia* 27:67–81.

Hadfield, M. G., S. E. Miller, and A. H. Carwile. 1993. The decimation of endemic Hawai'ian tree snails by alien predators. *Amer. Zool.* 33:610–622.

Hadfield, M. G., and B. S. Mountain. 1980. A field study of a vanishing species, *Achatinella mustelina* (Gastropoda, Pulmonata), in the Waianae Mountains of Oahu. *Pacific Science* 34:345–358.

Mead, R. R. 1979. Economic malacology with particular reference to *Achatina fulica.* In R. Fretter and J. Peake, eds., *Pulmonates.* New York: Academic Press.

Pilsbry, H. A., and C. M. Cooke. 1912–1914. *Manual of conchology, structural and systematic,* ser. 2, vol. 22, *Achatinellidae.* Philadelphia: Academy of Natural Sciences of Philadelphia.

Reif, W.-E. 1985. The work of John Th. Gulick (1832–1923): Hawaiian snails and a concept of population genetics. *Z. f. zool. Systematik u. Evolutionsforschung* 23:161–171.

Chapter 3: As the Hawaiian Crow Flies . . .

Audubon Society of Hawaii. 1989. *'Alala* (Hawaiian crow) recovery: Audubon Hawai'i calls for immediate government action. *Greenprint* 1 (2):1.

Committee on Scientific Bases for the Preservation of the Hawaiian Crow, Board on Biology, Commission on Life Sciences, National Research Council. 1991. *Scientific bases for the preservation of the Hawaiian crow.* Washington, D.C.: National Academy Press.

Harrington, C. Y., and G. Pinholster. 1992. Swift action urged to save Hawaiian crow. *National Research Council NewsReport* 42:12–14.

TenBruggencate, Jan. 1981. A controversial battle to save Hawaii's crows. *Honolulu Advertiser,* April 7.

———. 1989a. McCandless wild crows thriving. *Honolulu Advertiser,* Dec. 10.

———. 1989b. Ranch vetoes plan to corral last 'Alala. *Honolulu Advertiser,* May 5.

———. 1992a. Allowed 'Alala survey to avoid court, rancher says. *Honolulu Advertiser,* April 6.

———. 1992b. First goal is to increase 'Alala, national panel says. *Honolulu Advertiser,* May 7.

Chapter 4: Darwin, Ulysse, Rousseau—and Brutus, Too?

Boesch, C. 1994. Chimpanzees–red colobus monkeys: A predator-prey system. *Anim. Behav.* 47:1135–1148.

Boesch, C., and H. Boesch. 1989. Hunting behavior of wild chimpanzees in the Taï National Park. *Am. J. Phys. Anthrop.* 78:547–573.

Gagneux, P., D. S. Woodruff, and C. Boesch. 1997. Furtive mating in female chimpanzees. *Nature* 387:358–359.

Le Guenno, B., et al. 1995. Isolation and partial characterisation of a new strain of Ebola virus. *Lancet* 345:1271–1274.

Morell, V. 1995. Chimpanzee outbreak heats up search for Ebola origin. *Science* 268:974–975.

Chapter 5: Britain's Large Blue

Thomas, J. A. 1986. *Butterflies of the British Isles.* London: Reed Consumer Books.

———. 1995. Return of a native. *Natural World* 44:26–28.

Chapter 6: Monachus Monachus, in Retreat

Jacobs, J., and A. Panou. 1988. Conservation of the Mediterranean monk seal, *Monachus monachus*, in Kefalonia, Ithaca, and Lefkada Islands, Ionian Sea, Greece. Institut Royal des Sciences Naturelles de Belgique, Project ACE 6611/28.

———. 1996. Final progress report, WWF Project GR0034.01, Conservation Programme for the Ionian, Activity 1: Kefalonia and Ithaca.

Panou, A., ed. 1996. Monk seal conservation in Greece. Part 3, Central Ionian Sea. Final report to the European Commission, contract no. B4-3040/95/009/AO/D2.

Panou, A., J. Jacobs, and D. Panos. 1993. The endangered Mediterranean monk seal *Monachus monachus* in the Ionian Sea, Greece. *Biol. Conserv.* 64:129–140.

Scoullas, M., M. Mantzara, and V. Constantianos. 1994. *The book directory for the Mediterranean monk seal (Monachus monachus) in Greece.* Athens: Akti Pbl. Nikos Hatzopoulos.

Chapter 7: Introducing the Nile Perch

Acere, T. O. 1988. The controversy over Nile perch, *Lates niloticus,* in Lake Victoria, East Africa. *Naga, the ICLARM Quarterly* 11: 3–5.

Achieng, A. P. 1990. The impact of the introduction of Nile perch, *Lates niloticus* (L.), on the fisheries of Lake Victoria. *Journal of Fish Biology* 37 (supplement A): 17–23.

Anderson, A. M. 1961. Further observations concerning the proposed introduction of Nile perch into Lake Victoria. *East African Agricultural and Forestry Journal* 26:195–201.

Barel, C. D. N., ed. 1986. The decline of Lake Victoria's cichlid species flock. Zoologisch Laboratorium, University of Leiden, The Netherlands.

Fryer, G. 1960. Concerning the proposed introduction of Nile perch into Lake Victoria. *East African Agricultural Journal* 25:267–270.

Fryer, G. 1973. The Lake Victoria fisheries: Some facts and fallacies. *Biological Conservation* 5:304–308.

Gee, J. M. 1964. Nile perch investigation. East African Fisheries and Forestry Research Organisation. Annual Report, 1962–1963, pp. 24–26.

Goldschmidt, T. 1996. *Darwin's dreampond: Drama in Lake Victoria.* Cambridge, Mass.: MIT Press.

Goldschmidt, T., F. Witte, and J. Wanink. 1993. Cascading effects of the introduced Nile perch on the detritivorous/phytoplanktivorous species in the sublittoral areas of Lake Victoria. *Conservation Biology* 7:686–700.

Graham, M. 1929. The Victoria Nyanza and its fisheries: A report on the fishing survey of Lake Victoria, 1927–28. London: Crown Agents for the Colonies.

Ogutu-Ohwayo, R. 1992. The purpose, costs, and benefits of fish introductions: With specific reference to the Great Lakes of Africa. *Mitt. Internat. Verein. Limnol.* 23:37–44.

Okaronon, J. O. 1994. Current composition, distribution, and relative abundance of the fish stocks of Lake Victoria, Uganda. *African Journal of Tropical Hydrobiology and Fisheries* 5:89–100.

Seehausen, O. 1996. *Lake Victoria rock cichlids.* Zevenhuizen, The Netherlands: Verduyn Cichlids.

Seehausen, O., J. J. M. van Alphen, and F. Witte. 1997. Cichlid fish diversity threatened by eutrophication that curbs sexual selection. *Science* 277:1808–1811.

Stoneman, J., K. B. Meecham, and A. J. Mathotho. 1973. Africa's great lakes and their fisheries potential. *Biological Conservation* 5:299–302.

Wandera, S. B., and J. H. Wanink. 1995. Growth and mortality of *Dagaa (Rastrineobola argentea*, Fam. Cyprinidae) in Lake Victoria. *Naga, the ICLARM Quarterly* 18:42–45.

Witte, F., et al. 1992. The destruction of an endemic species flock: Quantitative data on the decline of the haplochromine cichlids of Lake Victoria. *Environmental Biology of Fishes* 34:1–28.

Witte, F., et al. 1992. Species extinction and concomitant ecological changes in Lake Victoria. *Netherlands Journal of Zoology* 42:214–232.

Witte, F., T. Goldschmidt, and J. H. Wanink. 1995. Dynamics of the haplochromine cichlid fauna and other ecological changes in the Mwanza Gulf of Lake Victoria. In T. J. Pitcher and P. J. B. Hart, eds., *The impact of species changes in African lakes.* London: Chapman & Hall.

Chapter 8: A Wild Dog Story

Burrows, R. 1992. Rabies in wild dogs. *Nature* 359:277.

Burrows, R., H. Hofer, and M. East. 1994. Demography, extinction, and intervention in a small population: The case of the Serengeti wild dogs. *Proc. Roy. Soc. Lond. B* 256:281–292.

———. 1995. Population dynamics, intervention, and survival in African wild dogs (*Lycaon pictus*). *Proc. Roy. Soc. Lond. B* 262:235–245.

Creel, S. 1992. Cause of wild dog deaths. *Nature* 360:633.

———. Conserving wild dogs. *Trends in Ecology and Evolution* 11:337.

Dye, C. 1996. Serengeti wild dogs: What really happened? *Trends in Ecology and Evolution* 11:188–189.

East, M., and H. Hofer. 1996. Wild dogs in the Serengeti ecosystem: What really happened. *Trends in Ecology and Evolution* 11:509.

Estes, R. C. 1991. *The behavior guide to African mammals*. Berkeley: University of California Press.

Gascoyne, S. C., M. K. Laurenson, S. Lelo, and M. Borner. 1993. Rabies in African wild dogs (*Lycaon pictus*) in the Serengeti region, Tanzania. *J. Wildlife Diseases* 29:396–402.

Ginsberg, J. R., G. M. Mace, and S. Albon. 1995. Local extinction in a small and declining population: Wild dogs in the Serengeti. *Proc. Roy. Soc. Lond. B* 262:221–228.

Heinsohn, R. 1992. When conservation goes to the dogs. *Trends in Ecology and Evolution* 7:214–215.

Kühme, W. 1965. Communal food distribution and division of labour in African hunting dogs. *Nature* 205:443–444.

Macdonald, D. W. 1992. Cause of wild dog deaths. *Nature* 360:633–634.

Morell, V. 1995. Dogfight erupts over animal studies in the Serengeti. *Science* 270:1302–1303.

Scott, J. 1991. *Painted wolves: Wild dogs of the Serengeti-Mara*. London: Hamish Hamilton.

Sinclair, A. R. E., and P. Arcese, eds. 1995. *Serengeti II*. Chicago: University of Chicago Press.

Chapter 9: SOS for a Salamander and Its Springs

Bruce, R. Don't say 'salamander.' *Austin Chronicle*, May 17, 1996.

Chippindale, P. T., A. H. Price, and D. M. Hillis. 1993. A new species of perennibranchiate salamander (*Eurycea*: Plethodontidae) from Austin, Texas. *Herpetologica* 49:248–259.

Kirkpatrick, M. 1996. Salamander trapped in collision between science and politics. *Society of Conservation Biology Newsletter*.

Pipkin, T. and M. Frech, eds. *Barton Springs eternal: The soul of a city*. Austin, Tex.: Softshoe Publishing, Hill Country Foundation.

Wright, S. Austin leaps to 8th place in magazine's city ratings. *Austin American Statesman*, June 13, 1996.

Chapter 10: Islands of the Living Dead

Ziswiler, V. 1996. *Der Dodo: Fantasien und Fakten zu einem verschwundenen Vogel*. Zurich: Zoologisches Museum der Universität.

Chapter 11: The Echo of Their Song

Watson, B. 1968. *The complete works of Chuang Tzu*. New York: Columbia University Press, p. 216.

Van Gulik, R. H. 1967. *The gibbon in China: An essay in Chinese animal lore*. Leiden: Brill.

Chapter 12: Watching, from the Edge of Extinction

Merton, T. 1965. *The Way of Chuang Tzu*. New York: New Directions.

INDEX

Chimpanzees: cooperative hunting in, 59, 67; Ebola, 64; habituation, 60, 72; related to humans, 74; in Taï, 56–58, 68; and tool use, 59, 60
Chuang Tzu, 229, 247
Comanche Springs, 197
Commission for Science and Technology, Tanzania. *See* COSTECH
Conservation: ethics, 244; movement, 175; politics of biology, 176; as trendy career, 243
Cook, Captain, 6, 8, 31
Copepod, 186, 240, 241
Corvus hawaiiensis. See *'Alala*
COSTECH, 151, 154, 168, 171

Dark-rumped Petrel, 9, 10
Darwin, Charles, 27, 65
Darwin (chimp), 62, 65
Development in Austin, money for, 189, 196, 205, 213, 239
Distemper, canine, in Namibian wild dogs, 184, 185
Diversity, in haplochromines, 137, 138
DNA: in birds, 15, 49; in chimpanzees, 58, 59, 72–74, 78, 79; in salamanders, 191
Dodo, 13, 16, 17, 214, 217
Duvall, Fern, 45–47, 50

East, Marion, 147, 167, 172, 176–187
Easter Island, 8
Ebola virus, 58, 62–65; origin of, 64, 65
Eco-mimics, 244
Ecosystem: collapse, 19; disruption, in Lake Victoria, 128, 139, 142; on Easter Island, 8; on Hawaii, 9; in Lake Constance, 241; native, 23; value of, 249
Ecotourism: Greece, 107; Hawaii, 54
Edwards Aquifer, 190, 194, 202, 207, 209–211, 213
Elster, Hans-Joachim, 240–242, 247
Elvis species, 39
Endangered Species Act: and Barton Springs salamander, 190, 201–204, 208; and Hawaiian crow, 47, 48
Endangered Species List, 20, 190, 202

Euglandina rosea, 28, 30, 32, 33, 38
European Community, 100, 102, 107
European Union, 108–110, 112
Eurycea sosorum. See Barton Springs salamander
Extinction: avoidance of, 49; causes, 11; crisis, 14, 37; drama, 239; ecosystem impact, 13; economists' perspective on, 4, 249; effect on history and culture, 246; of gibbon, 238, 248, 254; of haplochromines, 125–130, 132, 133, 138, 139, 141, 242; as holocaust, 240, 242, 243; imminent danger of, 198; on Kauai, 21; of large blue butterfly, 90–93; local, in wild dogs, 147, 153, 157, 158, 162, 165, 166, 168, 171, 172, 185; on Mauritius, 4, 12; meaning of, 3, 13, 35, 54, 62, 111, 132, 133, 139, 141, 142, 145, 173, 186, 195, 212, 242; through overfishing, 120, 123; prehuman, 6; as unimportant, 240, 241

Fisheries Research Institute, Uganda (FIRI), 140, 142–145
Fishermen, 101, 102, 105, 112; compensation for, 103, 106; hostility to seals, 98, 99, 103, 106, 113; letters to E.C., 104, 108
Flagship species, 174
Fossey, Dian, 61, 77
Frankfurt Zoological Society (FZS), 148, 151, 154, 159, 170, 177
Frauenschuh. See Lady's slipper
Freeport-McMoRan, 198, 200, 205, 206
Fryer, Geoffrey, 124

Gagneux, Pascal, 72–78
Gascoyne, Sarah (Cleaveland), 161, 162, 164, 165, 167, 171; at Arusha conference, 183, 184
Gibbon: chain-drinking, 230, 231; cultural role of, 237, 246; extinction of, 238, 248, 254; history of, 238; paintings of, 234, 236; as pet, 228; range of, 237; as symbol, 229, 230
Goodall, Jane, 59, 61, 179